# EXPLORING OPPORTUNITIES FOR
# STEM TEACHER LEADERSHIP
## Summary of a Convocation

Steve Olson and Jay Labov, *Rapporteurs*

Planning Commmittee on Exploring Opportunities for
STEM Teacher Leadership

Teacher Advisory Council

Division of Behavioral and Social Sciences and Education

**NATIONAL RESEARCH COUNCIL**
*OF THE NATIONAL ACADEMIES*

THE NATIONAL ACADEMIES PRESS
Washington, D.C.
**www.nap.edu**

**THE NATIONAL ACADEMIES PRESS**     500 Fifth Street, NW     Washington, DC 20001

NOTICE: The project that is the subject of this report was approved by the Governing Board of the National Research Council, whose members are drawn from the councils of the National Academy of Sciences, the National Academy of Engineering, and the Institute of Medicine.

This project was supported by the National Science Foundation (Award# DRL-1406780). Any opinions, findings, conclusions, or recommendations expressed in this publication are those of the authors and do not necessarily reflect the views of the organizations or agencies that provided support for the project.

International Standard Book Number-13:   978-0-309-31456-5
International Standard Book Number-10:   0-309-31456-9

Additional copies of this report are available for sale from the National Academies Press, 500 Fifth Street, NW, Keck 360, Washington, DC 20001; (800) 624-6242 or (202) 334-3313; http://www.nap.edu.

Printed and bound in Great Britain by Marston Book Services Ltd, Oxfordshire

Suggested citation: National Research Council. (2014). *Exploring Opportunities for STEM Teacher Leadership: Summary of a Convocation.* S. Olson and J. Labov, Rapporteurs. Planning Committee on Exploring Opportunities for STEM Teacher Leadership: Summary of a Convocation, Teacher Advisory Council, Division of Behavioral and Social Sciences and Education. Washington, DC: The National Academies Press.

# THE NATIONAL ACADEMIES
*Advisers to the Nation on Science, Engineering, and Medicine*

The **National Academy of Sciences** is a private, nonprofit, self-perpetuating society of distinguished scholars engaged in scientific and engineering research, dedicated to the furtherance of science and technology and to their use for the general welfare. Upon the authority of the charter granted to it by the Congress in 1863, the Academy has a mandate that requires it to advise the federal government on scientific and technical matters. Dr. Ralph J. Cicerone is president of the National Academy of Sciences.

The **National Academy of Engineering** was established in 1964, under the charter of the National Academy of Sciences, as a parallel organization of outstanding engineers. It is autonomous in its administration and in the selection of its members, sharing with the National Academy of Sciences the responsibility for advising the federal government. The National Academy of Engineering also sponsors engineering programs aimed at meeting national needs, encourages education and research, and recognizes the superior achievements of engineers. Dr. C. D. Mote, Jr., is president of the National Academy of Engineering.

The **Institute of Medicine** was established in 1970 by the National Academy of Sciences to secure the services of eminent members of appropriate professions in the examination of policy matters pertaining to the health of the public. The Institute acts under the responsibility given to the National Academy of Sciences by its congressional charter to be an adviser to the federal government and, upon its own initiative, to identify issues of medical care, research, and education. Dr. Victor J. Dzau is president of the Institute of Medicine.

The **National Research Council** was organized by the National Academy of Sciences in 1916 to associate the broad community of science and technology with the Academy's purposes of furthering knowledge and advising the federal government. Functioning in accordance with general policies determined by the Academy, the Council has become the principal operating agency of both the National Academy of Sciences and the National Academy of Engineering in providing services to the government, the public, and the scientific and engineering communities. The Council is administered jointly by both Academies and the Institute of Medicine. Dr. Ralph J. Cicerone and Dr. C. D. Mote, Jr., are chair and vice chair, respectively, of the National Research Council.

**www.national-academies.org**

## PLANNING COMMITTEE ON EXPLORING OPPORTUNITIES FOR STEM TEACHER LEADERSHIP

**MIKE TOWN** (*Chair*), Tesla STEM High School, Redmond, WA
**JANET ENGLISH,** El Toro High School, Lake Forest, CA
**CINDY HASSELBRING,** Maryland State Department of Education, Baltimore
**TOBY HORN,** Carnegie Institution for Science, Washington, DC
**SUSANNA LOEB,** Stanford University, Stanford, CA
**STEVE ROBINSON,** Democracy Prep Charter High School, New York

**JAY LABOV,** *Senior Advisor for Education and Communication*
**MARY ANN KASPER,** *Senior Project Assistant*
**MATTHEW LAMMERS,** *Program Coordinator*

# TEACHER ADVISORY COUNCIL

**STEVEN L. LONG** (*Chair*), Rogers High School, Rogers, AR
**JULIANA JONES** (*Vice Chair*), Longfellow Middle School, Berkeley, CA
**NANCY ARROYO,** Riverside High School, El Paso, TX
**CHARLENE DINDO,** Pelican's Nest Science Lab, Fairhope, AL
**KENNETH HUFF,** Mill Middle School, Williamsville, NY
**MARY MARGUERITE (MARGO) MURPHY,** Camden Hills Regional
   High School, Rockport, ME
**JENNIFER SINSEL,** Bostic Elementary School, Wichita, KS
**SHEIKISHA THOMAS,** Jordan High School, Durham, NC

**BRUCE ALBERTS** (*Ex Officio*), University of California, San Francisco

**JAY LABOV,** *Senior Advisor for Education and Communication and Staff
   Director*
**MARY ANN KASPER,** *Senior Program Assistant*
**MATTHEW LAMMERS,** *Program Coordinator*
**ELIZABETH CARVELLAS,** *Teacher Leader*

# Acknowledgments

This report has been reviewed in draft form by individuals chosen for their diverse perspectives and technical expertise, in accordance with procedures approved by the National Research Council's (NRC's) Report Review Committee. The purpose of this independent review is to provide candid and critical comments that will assist the institution in making its published report as sound as possible and to ensure that the report meets institutional standards for objectivity, evidence, and responsiveness to the study charge. The review comments and draft manuscript remain confidential to protect the integrity of the process. We wish to thank the following individuals for their review of this report: Bill Badders, Office of President, National Science Teachers Association, and science teacher, Cleveland Heights, Ohio; Sophia Gershman, teacher, Watchung Hills Regional High School, New Jersey; Kenneth Huff, science teacher, Williamsville Central School District, Williamsville, New York; Ken Krehbiel, Office of Associate Executive Director for Communications, National Council of Teachers of Mathematics, Reston, Virginia; Zovig Minassian, Albert Einstein Distinguished Educator Fellow, Office of Science Workforce Development for Teachers and Scientists, U.S. Department of Energy; and Terri M. Taylor, Office of Assistant Director, K-12 Education, American Chemical Society, Washington, DC.

Although the reviewers listed above have provided many constructive comments and suggestions, they did not see the final draft of the report before its release. The review of this report was overseen by Ford Morishita, science specialist at the Science and Mathematics Education

Resource Center in Vancouver, Washington. Appointed by NRC's Division of Behavioral and Social Sciences and Education, he was responsible for making certain that an independent examination of this report was carried out in accordance with institutional procedures and that all review comments were carefully considered. Responsibility for the final content of this report rests entirely with the authors and the institution.

Special thanks and deep appreciation are extended to Janice Earle in the National Science Foundation's Division on Research and Learning (Directorate on Education and Human Resources) for her many years of support and encouragement of work to improve science, technology, engineering, and mathematics (STEM) education at NRC and throughout the education system. Her commitment to exploring the evidence base for improving teaching and learning in STEM and her sheer hard work in helping to move forward effective research methods and practices are recognized and greatly appreciated by NRC and colleagues throughout the STEM education community.

Margo Murphy, *Chair* (as of July 1, 2014)
Jay Labov, *Staff Director*
Teacher Advisory Council

# Contents

# 1

# Introduction and Themes of the Convocation

Many national initiatives in K-12 science, technology, engineering, and mathematics (STEM) education have emphasized the connections between teachers and improved student learning. Much of the discussion surrounding these initiatives has focused on the preparation, professional development, evaluation, compensation, and career advancement of teachers. Yet one critical set of voices has been largely missing from this discussion— that of classroom teachers themselves (Berry, 2011).

Isolated examples of involving teachers in education policy and decision making have occurred at all levels of the education system (see Chapter 3; also Pennington, 2013). In addition, a number of studies have demonstrated that when teachers are effectively engaged in policy and decision making, teacher morale improves, retention may increase, and the school and surrounding communities benefit (Bhatt and Behrstock, 2010; Bissaker and Heath, 2005; Kimmelman, 2010; Rasberry and Mahajan, 2008). Given the benefits of involving a diverse group of teacher leaders in education policy and decision making, organizations have been seeking to empower teacher leadership at national, state, and local levels. For example, the Center for Teaching Quality, headquartered in North Carolina, is seeking to create schools "where America's most accomplished teachers routinely spread their expertise, enforce standards of teaching excellence, transform teacher preparation and certification, and redesign and lead their own schools."[1] A recently published report from this organization

---

[1] Additional information is available at http://www.teachingquality.org [September 2014].

advocates for "professional discussion about what constitutes the full range of competencies that teacher leaders possess and how this form of leadership can be distinguished from, but work in tandem with, formal administrative leadership roles to support good teaching and promote student learning" (Teacher Leadership Exploratory Consortium, 2011).

To explore the potential for STEM teacher leaders to improve student learning through involvement in education policy and decision making, the National Research Council (NRC) held a convocation on June 5–6, 2014, in Washington, DC, entitled "One Year After *Science's* Grand Challenges in Education: Professional Leadership of STEM Teachers Through Education Policy and Decision Making." The convocation was organized by a committee of experts under the aegis of the National Academies Teacher Advisory Council, which was established in 2002 to help bring master teachers' "wisdom of practice" to Academy staff and other organizations as they work to improve STEM education for grades K-12.[2] It was structured around a special issue of *Science* magazine that discussed 20 grand challenges in science education (Alberts, 2013). The authors of three major articles in that issue[3]—along with Dr. Bruce Alberts, *Science's* editor-in-chief at the time—spoke at the convocation, updating their earlier observations and applying them directly to the issue of STEM teacher leadership.

The Statement of Task for this project is as follows:

An ad hoc committee under the aegis of the National Academies Teacher Advisory Council will organize and conduct a two-day convocation in Washington, DC early in 2014 that would focus on empowering teachers to play greater leadership roles in education policy and decision making in science, mathematics, engineering, and technology (STEM) education at the national, state, and local levels. The convocation will use several papers from the April 2013 special issue of *Science* on Grand Challenges in Education as foci for discussion. The convocation will feature invited presentations and discussion that would explore the following issues:

- What is the evidence base (both from the United States and from other countries) that addresses whether involving teachers of STEM in education policy and decision making can lead to improvements in policies that affect teachers and teaching in these subject areas? What other evidence exists for improving other aspects of the K-12 STEM education system, especially at the national level?

---

[2]Additional information is available at http://nas.edu/tac [September 2014].
[3]Barnett Berry, Center for Teaching Quality; Suzanne Wilson, University of Connecticut, Storrs; and Suzanne Donovan, Strategic Education Research Partnership.

- What models of engaging teachers in policy and decision making at the national, state, and local levels currently exist, especially as they pertain to STEM education? What can be learned from current efforts to expand the roles of teachers in these processes?
- What kinds of communication efforts, resources and other activities are needed to help education officials and policy makers understand the roles, contributions and potential impact of teachers of STEM in these processes? The convocation will invite participation from a broad array of stakeholders from the STEM education and policy communities. Following the convocation a National Academies workshop summary (Category B) will be prepared by a rapporteur and distributed widely.

The convocation brought together representatives from a wide range of public- and private-sector organizations, including foundations and other funding organizations, for a day and a half of discussions, presentations, and breakout sessions. Convocation participants also included many teachers who have had leadership positions and who have been involved in programs to foster teacher leaders. The convocation allowed participation in person or via a live Webcast. (Appendix B lists in-person participants and registrants for the Webcast.) During the registration process, people were asked to (1) indicate their reason(s) for wanting to participate in this convocation, and (2) indicate what they hoped to take away from this convocation. The summary of responses that were presented during opening remarks is found in Box 1-1. In addition, a Twitter feed permitted all participants to provide real-time comments and feedback.[4]

During the opening session of the convocation, Mike Town, chair of the organizing committee and a teacher at Tesla/STEM High School in Redmond, Washington, said, "This is the first time we've had an opportunity to have a dialogue about policy and teacher empowerment with all the different stakeholders in the same room at the same time."[5]

The overall objective of the convocation, said Richard Duschl, a senior advisor at the National Science Foundation (NSF), which supported the convocation, was to discuss how teachers could "become leaders, become spokespersons, become individuals who know how to speak about policy and about the impact of the various evidence-based practices that we are now trying to put into place."

---

[4]The Twitter feed is available at #STEMTeachers. Participants could tweet at @TAC_of_NRC.

[5]Despite efforts to invite them, school administrators such as principals were missing from the stakeholders who attended the convocation. Some invitees responded that because the convocation was so close to the end of the school year and end-of-year assessments, their responsibilities would not permit them to participate.

---

**BOX 1-1**
**Convocation Survey Summary**

*Briefly, please indicate your reason(s) for wanting to participate in this convocation.*

| | |
|---|---|
| Gain the latest information regarding STEM education. | 35% |
| Learn how to best empower teachers in STEM education policy-making roles. | 27% |
| Participate in a dialogue on STEM research and integration. | 26% |
| Are members of the National Academies Teacher Advisory Council. | 7% |
| Networking opportunities. | 2% |
| Interested as NRC staff members. | 2% |
| Attending as a representative of a funding organization. | 1% |

*Briefly, please indicate what you hope to take away from this convocation.*

| | |
|---|---|
| Gain a better understanding of how they can empower teachers as STEM education policy makers. | 32% |
| Participate in a dialogue on how they could improve STEM education. | 23% |
| Gain an understanding about the future of STEM education policy. | 22% |
| Learn about the evidence concerning teacher empowerment in decision making for STEM education. | 11% |
| Network with colleagues. | 8% |
| Support the National Academies Teacher Advisory Council or the *Next Generation Science Standards*. | 4% |

---

The convocation also built on a meeting held in February 2014 (National Research Council, 2014) and a workshop held three days prior to this convocation[6] on how schools, afterschool programs, and institutions such as museums that provide informal STEM learning can work together to improve STEM education. As Jay Labov, NRC senior advisor for education and communication and the organizer of both events, posed, "How do we literally think about the learner rather than the institutions and obliterate, at least figuratively if not literally, the walls of the schoolhouse so that STEM learning can take place in all different places?"

This report summarizes the presentations and discussions at the convocation as a possible guide to future discussion and action. The report is

---

[6]Additional information about this workshop, "Successful Out-of-School STEM Learning: A Consensus Study," held June 3-4, 2014, under the NRC Board on Science Education, is available at http://sites.nationalacademies.org/DBASSE/BOSE/CurrentProjects/DBASSE_086842 [September 2014].

written as a narrative that highlights the important themes, opportunities, and challenges discussed by presenters and participants, rather than a chronological summary of the convocation. The observations and suggested actions in this report are those of individual speakers and should not be interpreted as the conclusions of the convocation participants as a whole. However, they provide an overview of the current situation and may point in promising directions. According to Labov, "We see this convocation not as a culminating experience but as the beginning of a process to think about how educators, whether formal, informal, or afterschool, can have their voices heard."

## THEMES OF THE CONVOCATION

At the beginning and end of the convocation's second day, members of the organizing committee, reporters for the breakout groups, and other convocation participants offered their reflections on themes they heard emerging from the presentations and discussions. Those observations are summarized here as an introduction to the major issues that arose at the convocation.

According to several participants, many opportunities exist for STEM teacher leaders to influence practices within the classroom and policies that shape the broader educational context. But the term "teacher leader" can be distrusted in the context of schools, where teachers typically work among themselves as equals rather than in a hierarchy, observed Toby Horn, codirector of the Carnegie Academy for Science Education at the Carnegie Institution for Science in Washington, DC. In addition, she said that teacher leadership can be viewed by administrators as taking time away from what is viewed as teachers' primary (or only) role of being in the classroom, working directly with students.

However, a variety of models around the country are demonstrating the value of involving STEM teachers in education policy and decision making. This leadership can make a difference both in individual classrooms and in policies that affect many classrooms, as several presenters pointed out. As an example, Town cited the opportunity for teacher leaders to influence professional development. Teachers know what kinds of professional development work in their classrooms. They recall professional development opportunities from which they became more confident, gained a stronger identity in STEM subjects, and learned new strategies to teach students. STEM teacher leaders can have an influence on professional development in their schools, districts, states, and nationwide, he suggested. For example, they have opportunities to identify inadequate professional development and replace it with something else. He suggested that administrators need to trust that teachers know best what their students need and give teachers the freedom to try new and exciting things. "That's a very effective way for us to use our voice," Town said.

Steve Robinson, a teacher at Democracy Prep Charter High School in New York City and a former special assistant for education in the Executive Office of the President of the United States, was the next respondent. He focused his remarks on accountability, reminding convocation participants that this issue will not disappear. Federal and state policy generally includes leeway for educators to devise alternate measures of teacher effectiveness. However, Robinson noted that until people in the field themselves can develop other measures that are determined to be valid and reliable measures of what an effective science teacher looks like, policy makers will say, 'I'm going to use the numbers that come from tests, because that's the simplest thing to do.' Defining teacher effectiveness is complicated, but it is a challenge that we have to face," Robinson said.

Creating leadership opportunities for STEM teachers requires better strategies for communicating with principals, Horn noted. Principals need to know that STEM subjects are different than other subjects, she said. They also need to know that everyone can learn STEM subjects, not just some people, and that STEM learning is for all students from elementary school to high school. Teacher leaders can help principals see and understand these attributes of STEM teaching, which in turn can help principals foster such change in their schools and make change sustainable, she said.

STEM teacher leaders also need to be recognized and rewarded in some way, noted Cindy Hasselbring, special assistant to the state superintendent for special projects at the Maryland State Department of Education. STEM teacher leaders cannot be expected to do all of their leadership work on weekends and in the evenings. Acknowledging and promoting teacher leadership also requires that they have time during their workdays outside the classroom. They need the flexibility to connect with each other, strengthen local opportunities, and build communities of teachers as learners, which implies the existence of reward systems established between teacher leaders and administrators, Hasselbring said.

Teachers will make different decisions regarding leadership, noted Janet English, a high school science teacher in California who recently received a Fulbright Fellowship to study the Finnish education system.[7] Some will decide on being involved at different levels, from the school level to the national or international level while others will not—perhaps because they do not want to leave the classroom or are worried about job security. For some teachers who are reticent, mentoring may convince

---

[7]Additional information about the Fulbright Distinguished Awards in Teaching Program is available at http://www.iie.org/Programs/Fulbright-Awards-In-Teaching [September 2014].

some of them to become more involved in leadership activities, English said.

Teacher leaders are not guaranteed acceptance or even understanding by their administration of the value that they add to the classroom, school, or district, English added. Financial considerations are a factor for teachers if they leave the classroom for any substantial period of time. In some cases, she commented, "the people who do this are people who are willing to give up on a lot to get a lot. It's a tradeoff, but it's not generally easy to come out of the classroom and do something else," English emphasized.

Finally, English noted that, in other countries, teachers are involved not just in working with children, but also in their own development as teachers and professionals. They work together and give each other feedback on what helps children learn. They collaborate to create effective educational environments and to gain the respect of people who make policy decisions. (See also Organisation for Economic Co-operation and Development, 2011.)

## SUGGESTIONS FROM BREAKOUT GROUPS

On the first day of the convocation, participants separated into breakout groups to consider more detailed questions and options for future action based on the principles discussed by the plenary speakers that morning. Each presenter participated in a breakout session twice that afternoon, and each participant was able to attend two different sessions (details are in Appendix A). This first set of breakout groups looked at (1) professional development to engage teachers in leadership activities (the subject of Chapter 4), (2) the concept of "teacherpreneurs" (discussed in more detail in Chapter 2), and (3) current models of engaging teachers in leadership activities (some of which are described in Chapter 3). The breakout groups on the second day, which were organized by sector, suggested actions that could be taken in the short term and medium term. The following summary of conclusions and proposals from the breakout groups should not be viewed as representing a consensus of either the breakout groups or the convocation participants, but they raise intriguing issues and possibilities.

### Developing a Tool for Collaboration

Teacher leaders will create leadership in their own contexts, said English, who reported out for the group on models of engaging teachers in leadership activities. This means that every teacher leader will be different, because every context is different. A commitment to lead, therefore, will take different forms depending on what is appropriate and needed in an individual school, district, or state.

Nevertheless, English continued, teacher leaders need a way to communicate their experiences, learn from each other, and develop best practices. In the past, experienced and dedicated teacher leaders have not necessarily had a mechanism that enabled them to collaborate, she pointed out. Funders, either within government or outside, could help create such a mechanism. "Without a collaborative tool, there is no way to talk among ourselves," said English. "We need money and a commitment by funders to say, 'We recognize you as professionals, and we recognize that in order to improve education we need the teacher voice.'"

The groundwork for such collaboration already has been laid, added Robinson, by programs to recognize and reward especially effective STEM teachers, such as the Presidential Awards for Excellence in Mathematics and Science Teaching.[8] A network of teachers recognized through such programs could draw on teachers who are not teaching full time to write the initial proposals and provide continuing impetus for the resource. "The network is going to need some external expertise, which is maybe funding or maybe people who are clever about how to build good websites and social networks," Robinson said. The network also could interact with professional societies, teachers unions, and other organizations to refine and pursue its mission, he noted.

Different programs have had different degrees of success in keeping people connected after an event or program, English observed, noting "you have to have something to keep people engaged and interested in being part of that group." A mechanism suggested by Alberts during the breakout session was a series of meetings patterned after the Gordon Research Conferences in science,[9] where people come together for one week a year and "the most interesting thing at the retreat is the other people who are there."

### Expanding on Models of Teacher Leadership

Rebecca Sansom, an Einstein fellow[10] at the National Science Foundation, elaborated on some of the ways in which existing models of STEM teacher leadership could be expanded. One option she suggested would be to provide teacher leaders with more opportunities for authentic research opportunities as a way of developing their leadership knowledge and

---

[8]Additional information is available at https://www.paemst.org/ [September 2014].

[9]Additional information is available at https://www.grc.org/ [September 2014].

[10]Additional information is available at http://www.trianglecoalition.org/einstein-fellows [September 2014]. This program is also described in greater detail in Chapter 3.

skills. These research experiences could play a role in teacher preparation programs and certification, which would help teachers better understand the nature of science and use that understanding in their classrooms.

Another possibility would be to conduct a survey to examine the attitudes and dispositions of teachers, administrators, and other educators about the value of teacher input to education policy and decision making. This survey could be conducted by a teacher organization such as the National Council of Teachers of Mathematics[11] or the National Science Teachers Association,[12] she said, or it could be added to other surveys of teachers being done currently.

An organization such as the National Academies could prepare an introduction to the issue of teacher leadership and a guide to action, she suggested. A regular publication could point teacher leaders to important recent publications. This information also could interest many other people in education, such as teacher educators, professional development providers, and administrators.

### Defining Roles and Policy Issues

Though teacher leaders will engage in different activities, some of the activities in which they are involved will fall into the same category, said Camsie McAdams, senior STEM advisor at the U.S. Department of Education. Therefore, another step forward would be to define roles of teacher leadership, such as instructional leader, professional development coach, mentor teacher, curriculum developer, adjunct professor, and teacher trainer. A separate workshop could define and develop these roles for teacher leaders, along with the professional development needed to prepare for those roles, McAdams said.

Such a convening also could define the areas of education and policy making in which teacher leaders could have a beneficial influence. Examples cited by McAdams were scheduling, staffing, teacher seminars, resource allocation, bond issues, state policy licensure, graduation requirements, credentialing, and certification. The discussion at the convocation focused on "what everyone is passionate about, which is instructional leadership," she observed, but many other policies could be influenced by teachers. Some of these policies, such as scheduling, may be subtle, but she said they make "a huge difference at the school level." For instance, scheduling can help determine whether teachers have an opportunity to talk together about best practices. This is a way for teachers to be "empowered at the building level," said McAdams.

---

[11]Additional information is available at http://nctm.org [September 2014].

[12]Additional information is available at http://nsta.org [September 2014].

## Professional Development for STEM Leadership

Professional development in STEM fields can be different than professional development in other fields because of the constantly changing base of knowledge in STEM areas, noted Town, who reported out for the breakout group on professional development. Furthermore, professional development itself changes over time as knowledge and practices improve. For example, a new and promising approach, said Town, is teacher-led professional development that results in portfolios demonstrating professional growth over time.

One challenge in developing leadership is measuring the effectiveness of professional development, he said, adding that the scores of students on standardized tests are not necessarily the only or most reliable measure of effectiveness. Others metrics might include such things as building confidence in teachers, getting students to take more science classes, graduation rates, and whether students pursue STEM subjects in college.

Professional development is required in many but not all professions, noted Town, which raises the issue of why professional development is considered such an important component of teaching. A related question is who sets the goals for professional development—teachers or some other group? Several members of the breakout group made the case that the best professional development comes from the bottom up, where teachers decide what is needed and agencies try to respond to those needs.

Professional development also can be oriented more toward policy making than toward teaching practices, he said. Policy changes can have a more widespread effect than classroom improvements if they are system-wide, which may argue for professional development focused on such changes, Town said. For example, professional development can be effective if it enables teachers to voice their concerns to policy makers at the federal level. However, policy makers tend to be more focused on accountability and data than are teachers, who instead tend to focus on best practices and students' experiences in the classroom.

Finally, Town directed attention to the changing roles of teachers, which also will require new forms of professional development. Despite increased understanding about effective approaches based on emerging evidence, changes to professional development also are influenced in part in response to the perceptions of policy makers. For example, he said, today those perceptions are very oriented toward accountability, which is driving professional development. Rebranding what professional development is or changing the perceptions of policy makers could both influence the professional development teachers receive.

### Funding STEM Teacher Leaders

Jo Anne Vasquez, vice president for educational practice for the Helios Education Foundation, reported back from the breakout group of funders. She said funders need to find a way to define teacher leadership in a way that is both coherent and accessible to them. Education issues are surrounded by a fair amount of "noise," she and several other convocation participants said. Funders need a way to separate critical issues in education from that noise, identify a high-priority issue, and focus their attention on ways to support those who are working to address that issue. One way to help do that is to have teachers integrally involved in writing grant proposals. "You cannot leave the teachers' voice out," she said.

Vasquez also pointed out that funders have different funding mechanisms. For example, the Helios Education Foundation does invited solicitations rather than requests for proposals. Networks of funders such as the STEM Funders Network—an affinity group that explores best practices in grant making for STEM activities—offer one way to reconcile different procedures and focus on promising areas. Interagency initiatives at the federal level and public-private funding partnerships are other ways of supporting activities and programs that could foster STEM teacher leaders, she said.

# 2

# What STEM Teacher Leaders Can Contribute to Education Policy and Decision Making

Several presenters at the convocation addressed the issue of what science, technology, engineering, and mathematics (STEM) teacher leaders can contribute to education policy and decision making. These contributions could take many different forms, depending on teachers' interests and the contexts in which they work. The result, as the speakers' presentations detailed, is a vast and largely untapped opportunity for STEM teacher leaders to improve student learning.

## THE CHALLENGE OF LEADERSHIP

The changes required to implement the Next Generation Science Standards (NGSS) require extensive involvement by teachers, noted Bruce Alberts, University of California San Francisco, in his introductory remarks at the convocation. The NGSS increasingly will require that students "use information—sort, analyze, and critique it—to make and defend arguments, solve problems, and incubate ideas," he said. Teachers, he noted, will be at the center of the synergistic changes required throughout K-12 education to achieve this vision.

More broadly, teachers are at the center of a complex and interconnected system, Alberts pointed out (see Figure 2-1). Though these interconnections provide opportunities to improve and support teaching—Alberts emphasized the role of college faculty in producing graduates who are prepared to become excellent STEM teachers and teacher leaders—they also can produce gridlock in attempting to change the system.

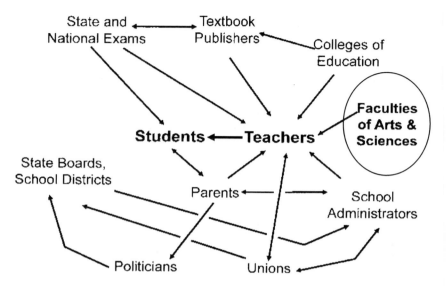

**FIGURE 2-1** Teachers are at the center of a network of influences, which can provide them with leverage for change but also yields a system prone to gridlock. SOURCE: Modified from National Research Council (1990, Fig. 1, p. 97).

As Alberts pointed out, schools remain predominantly hierarchical organizations. Even as businesses have learned to harvest "ground truth" from their employees to improve systems, many schools have remained relentlessly top-down. "What keeps me up at night is that our best teachers need to have much more influence on the education system at every level, from districts to states to the federal government," he said.

Alberts' desire to involve teachers in education policy and decision making was a major impetus behind the creation of the Teacher Advisory Council at the National Academies. A similar organization, modeled after this council, also exists in California and has been extremely successful in making the voices of teachers heard there. "We need one in every state," said Alberts, "and we need to empower them to have an effect on state policies."

Alberts also emphasized the need for funding organizations to support projects that replicate and adapt what works and not always focus on innovation. Funding agencies must work to decrease the strong incentives for "uniqueness," which is an enemy of coherence, he said. Government needs to support programs that have been shown to work so they are used much more widely. "If all we have is innovation and no spreading of what works, then we won't make much progress," he stated.

## THE BENEFITS OF TEACHER LEADERSHIP

"What are the benefits of engaging teachers in policy?" asked Diane Briars, president of the National Council of Teachers of Mathematics, who also was the mathematics director for the Pittsburgh Public Schools for 20 years. She provided two answers in her presentation.

When policy makers think about the best ways to attract and retain talented STEM teachers, they tend to think about pay raises, she said. But "that's typically not what I hear when I talk to teachers," said Briars. Instead, she related, teachers are more interested in things like teaching one less course so they have time during their workday to collaborate and engage in professional activities. Teacher engagement in policy making could help implement such ideas.

Teachers also can help change what they may see as counterproductive policies. For example, many teachers are overwhelmed with assessments at different levels, including district, state, and federal levels, said Briars. Involving teachers in education policy would help reveal how multiple policies coming to bear on a single classroom can have the opposite of the effect desired by the policy makers who design or implement them.

Teachers need time to be prepared and to become engaged with education policy and decision making, Briars noted. A major difficulty in education today is the public perception that teachers are working only when they are in front of students. Teachers need time to interact with other teachers, reflect, and improve their own practice, she said.

At the same time, Briars added, teachers have the most convincing voices on education issues of interest to the public, such as the implementation of the Common Core State Standards. Teachers need time and support to develop not only the knowledge but also the advocacy strategies to influence policy, she stated, "so that we are all working together to improve mathematics and science education and not working across purposes."

## CHEMISTRY TEACHERS AT THE AMERICAN CHEMICAL SOCIETY

High school chemistry teachers have played "a prominent role in advising and guiding the development, design, and implementation of a number of our programs, products, and service offerings," said Terri Taylor, assistant director of K-12 education at the American Chemical Society (ACS). As she explained, the ACS is a large scientific association, with more than 161,000 individual members. Its areas of strategic focus include providing information, advancing members' careers, communicating chemistry's value, and improving education, and it pursues these objectives through a robust network of members, committees, divisions, and local sections.

Over time, education has increased in priority within the ACS, said Taylor. The society has a Committee on Education, which has oversight for educational activities across the organization. One of the committee's activities has been to develop policy statements, such as the ACS policy statements "Science Education Policy" and the "Importance of Hands-on Laboratory Activities." The committee also is responsible for the development of guidelines documents, including *The ACS Guidelines and Recommendations for the Teaching of High School Chemistry* (American Chemical Society, 2012), and it reviews and responds to standards developed by others, such as the NGSS.

High school chemistry teachers have been heavily involved in all of these activities. For example, high school chemistry teachers were integral to the writing and dissemination of the first and second editions of *Chemistry in the National Science Education Standards* (American Chemical Society, 2008). They also are a key part of the development of resources that incorporate the NGSS. Other ACS publications, such as *ChemMatters* magazine and the textbook *Chemistry in the Community*, have benefited tremendously from the guidance and leadership of high school teachers, said Taylor. High school teachers were also directly involved in the advocacy for and establishment of a new organization at the ACS, the American Association of Chemistry Teachers, which was scheduled to launch shortly after the convocation.

High school chemistry teachers bring irreplaceable assets to partnerships with others, said Taylor, including expertise, practical experience, a diversity of perspectives, strategic direction, creativity, and a deep understanding of students, other teachers, and administrators. These attributes "impact our programs and our policies in very significant ways," she said.

Taylor also described several challenges in involving teachers in education policy and decision making. One issue is the best way to involve novice teachers. Achieving geographic and demographic diversity among teacher leaders also can be challenging, she said. Teachers need logistical and administrative support to participate in policy making. For example, the ACS often pays for substitutes to allow for teachers to travel to meetings. Also, some chemistry teachers resist identifying themselves with a disciplinary organization like the ACS. Taylor noted that sometimes there is a struggle to get educators to buy-in when the American Chemical Society tells them that they are chemistry teachers, that they belong, and that the ACS wants their input and leadership. Lack of this kind of self-identification can be a barrier to getting some great people in the room, said Taylor.

Finally, Taylor observed that she was a high school chemistry teacher who became involved with the ACS and eventually became a staff member. "It's a testament to the fact that this can work out well," she said.

## INVOLVING TEACHERS IN POLICY

Cindy Hasselbring, special assistant to the state superintendent for special projects at the Maryland State Department of Education, brought perspectives from two states where she has worked, Michigan and Maryland. In Michigan, where she was a mathematics teacher for 16 years, excellent teachers are recognized in various ways, such as through teacher-of-the-year awards at the state level and National Board Certification and Albert Einstein Distinguished Educator fellowships at the national level. In Michigan, a single person was in charge of several of these recognition programs, and this administrator realized that the teachers selected by these programs constituted a tremendous resource. She formed an organization called the Network of Michigan Educators that included teachers who had earned these various forms of recognition, with funding from the Michigan Department of Education.

These teachers "bring a different voice" to education policy, said Hasselbring. For example, they talk at state board of education meetings not about the negative aspects of teaching but about the many positive things that are happening in Michigan schools, such as innovative teaching practices and partnerships between schools and community organizations. For state board members, "that has become one of the favorite parts of the state board meeting," said Hasselbring. The network is also available to answer questions from the state superintendent or state legislative education committee about such issues as the Common Core State Standards, NGSS, or teacher certification. In addition, teachers belonging to the network provide leadership at state education summits. "They are used as a sounding board," said Hasselbring. "They are valued as a voice that's different."

In Maryland, recognized teachers often serve as providers of professional learning. Also, teachers-of-the-year meet with the superintendent to discuss how to address existing challenges, award-winning teachers are providing input to the STEM Education Strategic Plan on which Hasselbring is working, and every school has one trained STEM teacher leader. "People need to value the voice of a teacher," Hasselbring said. "If you have that, you can find ways to use them in policy."

## AVENUES OF TEACHER INVOLVEMENT

Francis Eberle, acting deputy executive director for the National Association of State Boards of Education, noted that many teachers want to work on change outside the classroom, not just among their students. But, he said, teachers take different avenues to work toward this goal.

One avenue, as noted by Hasselbring, is to meet with boards of education and other policy makers. "It provides a reality check for policy makers," said Eberle. As an example, Eberle mentioned the issue of graduation requirements. Such requirements can be easy to define for traditional sequences of science and mathematics classes, but the move to more integrated, technology-oriented, or competency-based sequences of classes raises questions about how to credit students for their achievements. Other policy issues where teachers could have valuable input include teacher licensure, partnering with colleges or universities, or implementing the NGSS. In all of these areas, said Eberle, "the teacher voice is very important."

Eberle also raised several challenges in getting teachers involved in education policy and decision making. Changing policy typically requires persistence, he said. It may take months to make a policy change, or it may require waiting for an entire election cycle. Also, teachers can work only part-time on an issue, whereas lobbyists and others can devote much more time to that cause.

A second challenge identified by Eberle is being prepared to advocate for and implement change. Policy making can be a complex process, involving public hearings, other forms of input, events that have to occur in sequence, and so on. To be effective, teachers need to know how the process works.

Finally, policy making often relies on networks, he noted. Teachers have networks of their own that they can use to influence policy, but these often are different than the networks of policy makers. Nevertheless, teachers' networks can be very influential in shaping certain kinds of policies, such as those involving student assessments, particularly when teachers are able to involve other organizations in pushing for a policy change.

## TEACHERPRENEURS FOR THE 21ST CENTURY

Barnett Berry, the founder, partner, and chief executive officer of the Center for Teaching Quality (CTQ), started his presentation with three pieces of evidence that make the case for teacher leadership.

Teaching remains one of the most revered professions in the eyes of the public, said Berry, but many members of the public still have a difficult time thinking that teachers are working in the best interest of their students when they are not in front of a classroom. However, no other top-performing nation in international comparisons of educational attainment has its teachers in front of a classroom for so many hours per week as in the United States, Berry said. Teacher schedules in other countries also differ from day to day, whereas in U.S. schools, most teachers have pretty much the same schedule day after day. Such flexibility launched and fuels the kind of creativity that is expressed in leadership, said Berry.

In addition, the top-performing countries in international comparisons of STEM student achievement, such as Finland and Japan, or provinces and cities, such as Shanghai and Singapore, invest heavily in teachers and teacher leadership. This teacher preparation and professional development has a heavy emphasis on developing research skills, which Berry said "is the root of much of the leadership that then emerges from those teachers once they enter the classroom." Teachers do lesson study[1] to become more expert in both content and pedagogy, and the professional development systems in these countries provide opportunities to get beyond "what" to teach to "how and why" to teach. For these and other reasons, teaching is more quality-based than data-driven in other countries.

Top-performing countries have a thinner curriculum and more flexibility at the school level for teachers and principals to design a schedule to meet the needs of students, Berry noted. Teachers also have the time to be influential in their communities and to build greater trust with the public so that there is less need for external accountability systems that can limit teacher effectiveness. These countries have sophisticated systems to evaluate teachers and identify teachers needing improvement. Also, Berry reported, in these other countries, 60 to 80 percent of credentialed educators teach children at least part of the day, whereas in the United States, the percentage ranges from 42 to 48 percent.

Polls show that teachers in the United States are looking for opportunities to lead that do not require them to leave teaching, Berry said. Most teachers—84 percent—say they are not interested in becoming a principal, according to one survey (MetLife Foundation, 2012). However, about half said that they are somewhat interested or very interested in hybrid roles that involve both teaching and leading.

CTQ has become an incubator of teacher leaders using this hybrid model, Berry said. It is a champion of "teacherpreneurs," which Berry defined as teachers who still teach regularly but have the time and incentives to execute their own ideas. More than 5,000 teachers are involved in the community of teacherpreneurs that Berry has helped organize. Some are being supported to teach half time and spend the rest of their time in leadership positions. Some are being funded as virtual community organizers for a variety of projects and initiatives. About 300 have been published in high-profile blogs associated with publications like *Education Week*, *Huffington Post*, and *Politico*. They also have written articles for peer-reviewed journals and four books using other teachers as sources and as peer editors. "If we want more teacher

---

[1]Chapter 4 of National Research Council (2010) describes the use of lesson study and other forms of professional development in China and compares those practices with typical U.S. practices.

leaders, we have to help more teachers, who are very busy, go public with their ideas," Berry said.

Communication is a critical component of these efforts, according to Berry. CTQ provides support to teachers to develop messaging and works with funding partners to elevate teacher voices. For example, a teacherpreneur in Florida had the idea to explain to the public what teaching is in 140 characters or less so the explanations can be sent out as text messages or Tweets. The resulting messages reached more than 3 million people. "This is our first foray in a campaign like this, and there is a lot that can be done with social media to help teachers cultivate their message," Berry said.[2]

CTQ also has worked closely with the National Education Association to support teacher leaders and help them rise to positions within the union and lead in structural reform. This is one way, said Berry, for the teacher voice to be front and center while the union voice is present as well.

Teachers have many potential opportunities to learn how to lead, Berry noted. They can mentor each other, engage in "externships" with organizations outside school (which tend to be shorter and more practice focused than internships), and test out their ideas and plans in safe places. They can talk about their ideas with members of the public and have the time to travel and listen to input from stakeholders.

Berry talked about removing some of the barriers to teacher leadership. Leadership requires preservice and professional development programs that both encourage and prepare teachers to lead. It thrives under administrators who understand the conditions needed for teachers to spread their expertise to each other. It requires state and local policies that encourage school districts to reallocate resources and rethink curriculum in ways that capitalize on teacher leadership. And it calls for evaluation and compensation systems that encourage teachers to lead and take risks. Berry suggested that a valuable federal initiative would be to provide incentives to create conditions under which teachers can lead but not leave. "Plenty of teachers are ready to lead if they had the opportunity and space to do so," he commented.

"We can do this," Berry concluded. "Hundreds of thousands of American teachers are ready to both teach and lead. With that type of leadership capacity, we can change the world."

---

[2]Additional information is available at http://www.teachingquality.org/teachingis [September 2014].

## COUNTERING INCREASED EXPENDITURES

During the discussion period, several presenters focused on ways to counter administrators' reluctance to promote teacher leaders because of the possibility of increased expenditures. Steve Long, a teacher at Rogers High School in Rogers, Arkansas, and chair of the National Academies Teacher Advisory Council, gave an example of the problem. Until recently he had been able to teach four or five periods a day instead of six and devote the rest of his time to leadership activities. But as funding tightened in recent years, the school informed him that he would need to teach six periods a day again. 'If you are not in the classroom, then we have to hire additional teachers, and it takes money to make that happen," he said he was told. Grants can help fill the gap, but any given grant will run out eventually.

Berry responded that other countries have a higher percentage of people in their school systems teaching classes, so each person can teach less. Other countries also have a thinner curriculum that can be organized more flexibly, which gives educators more space to organize their activities. A related initiative, which is happening in some parts of the United States, is to have teachers work on what are called "thin contracts," which do not specify exactly what a teacher should be doing. In this way, he said, unions can help teachers take leadership roles that go beyond their activities in the classroom.

In addition, Berry pointed out that new models of education involving cyberspace, team teaching, and flexible classrooms can help bring new approaches to scale. He also asked if unions could help teachers gain joint appointments at other institutions, including colleges and universities, as another avenue for teachers to provide critical input more routinely on issues such as preservice and in-service education.

Briars suggested addressing the issue from a research perspective. If policy research could demonstrate the benefits to students and schools of teachers' participation in decision making, administrators would recognize the multiple benefits of such involvement. They will realize that "this is something that's going to benefit the system overall . . . when we support teachers engaging in these kinds of professional activities beyond the classroom," she said.

Eberle added that collaboration among districts could reveal these benefits even more clearly by showing how teachers' activities have effects extending beyond the district. However, Robinson observed that enabling teachers to get involved in policy does sometimes mean that they leave a school, which is a loss to that school.

Janet English, steering committee member and teacher, pointed out that an especially productive way to encourage schools and districts to allow their teachers to take on leadership activities is to point out the ways in which those activities benefit schools and students. "I've always tried to define activities that benefit the district as well as the profession and me as a teacher," she said. She said she also looks for opportunities that come with funding so that the school does not have to cover her time out of the classroom.

Camsie McAdams, U.S. Department of Education, pointed out that the Teacher Incentive Funds made available by the department can support many of the activities discussed at the convocation, including time away from the classroom, different forms of compensation, different roles, and leadership opportunities.[3] She also observed that the department has recently launched a $35 million Teacher Quality Partnership Grant Program, which is designed to enhance the preparation of prospective teachers and the professional development activities for current teachers; hold teacher preparation programs at institutions of higher education accountable for preparing highly qualified teachers; and recruit effective individuals, including minorities and individuals from other occupations, into the teaching force.[4]

---

[3]More information is available at http://www2.ed.gov/programs/teacherincentive/index.html [September 2014].

[4]More information is available at http://www2.ed.gov/programs/tqpartnership/index.html [September 2014].

# 3

# Models for Engaging Teachers' Voices

Various models exist for involving teacher leaders in education policy and decision making (e.g., Pennington, 2013). Presenters at the convocation examined six in detail, and several others were mentioned. Some of these programs are well established, while others are forging new models for promoting teacher leadership and professional growth.

## ALBERT EINSTEIN DISTINGUISHED EDUCATOR FELLOWSHIPS

The Albert Einstein Distinguished Educator Fellowship Program began in 1990 with a grant from the John D. and Catherine T. MacArthur Foundation to the Triangle Coalition.[1] In 1994, the Albert Einstein Distinguished Educator Fellowship Act was signed into law, which gave the U.S. Department of Energy the responsibility for managing the program. The fellowship allows accomplished K-12 science, technology, engineering, and mathematics (STEM) educators to spend 11 months working in a federal agency or in a U.S. congressional office, bringing their extensive knowledge and experience in the classroom to education program or policy efforts. Fellows provide a teacher's perspective on federally funded programs and policy and contribute to the work of the individual offices in which they are placed, said Anthonette Peña, director of the program at the Triangle Coalition for STEM Education, which administers

---

[1]For more information, see http://www.trianglecoalition.org/einstein-fellows [September 2014].

the program. Currently, the program's sponsoring agencies include the U.S. Department of Energy, National Science Foundation (NSF), National Aeronautics and Space Administration, and National Oceanic and Atmospheric Administration, and four fellows are placed in House and Senate offices in the U.S. Congress each year.

Prospective fellows submit applications online in the fall, which are reviewed externally. The applications of the top competitors are sent to the participating federal agencies and sponsors, which do internal evaluations of the applicants and decide whom to bring in for face-to-face interviews. Interviews are generally done in February, with selections in the spring so that the selected fellows can prepare to move to Washington, DC, for the following school year.

The Triangle Coalition provides professional development for the fellows, Peña noted. They get together at least twice a month as a group and collaborate to learn about resources that are available locally through participating agencies and other organizations.

"It's amazing what these Einstein Fellows can accomplish within one year," said Peña. According to an evaluation done in 2012–2013, fellows contributed to the work of their offices, grew professionally, and gained resources on which to call in the future. Among the 26 fellows in 2013, 7 out of the 19 fellows who reported their activities following the fellowship returned to the classroom, 4 stayed for a second year, 6 took on a STEM leadership role, and 2 entered full-time doctoral programs. "The paths that Einstein fellows take after their fellowships vary significantly," said Peña. Some have to resign from their positions to take the fellowship, which can affect their subsequent choices. Others take a leave of absence and are committed to going back. The common denominator, said Peña, is that the fellowship is a catalyst for educators to move on to do "great things," whether in their original schools, at the district or state level, or at the national level.

## KNOWLES SCIENCE TEACHING FOUNDATION FELLOWSHIPS

The Knowles Science Teaching Foundation Fellowships are designed to improve STEM education by building a network of teachers who are also leaders in the classroom.[2] The program supports teachers to be "primary agents of change in the education system," said the program's director, Nicole Gillespie. It does this primarily through a five-year fellowship program for beginning high school STEM teachers. It also has a senior fellows program that sustains teachers as leaders and as stewards of the profession over the long term.

---

[2]For more information, see http://www.kstf.org [September 2014].

The program uses a hybrid model in which fellows come together at least three times a year and also participate in an online community where they can work together. Each teacher receives $4,000 per year for professional development of their choosing. Teachers also receive money for classroom materials and support to apply for National Board Certification and for leadership grants. "We are committed to being open about what teacher leadership looks like," said Gillespie. "Our baseline definition is that it has to have an impact outside your classroom."

More than 90 percent of fellows who begin the five-year program complete it. Of the teachers who have completed the program, more than 80 percent are still teaching in K-12 education. More than half of the others are at home caring for children; only three are not in education in some capacity.

The program especially focuses on leadership capacity. "We are deliberately trying to have a broad vision of what leadership means," said Gillespie.

More than 250 teachers have been fellows across 42 states, and one of the products of the program is the network of fellows. "We are starting to explore the idea of what it means to have a network of teachers. How does this help teachers? How does this help the system more broadly?" said Gillespie. For example, the program has recently encouraged the network to address issues of common interest and drive change from the bottom up.

An evaluation that interviewed the principals of fellows found them to be an informal resource to other teachers in the school or district, to lead or facilitate professional development workshops or seminars, to share ideas or resources, to lead or facilitate teacher study groups, to serve as a mentor or coach, and to observe and provide feedback to other teachers. The vast majority of administrators find the fellows "to be outstanding resources in their schools," Gillespie noted.

## MATH FOR AMERICA

Math for America (MfA) DC was launched in 2008 with the goal of ensuring excellence in mathematics teaching in Washington, DC, public secondary schools through the recruitment, training, and retention of talented mathematics teachers over a five-year period, said the program's coordinator, Marlena Jones.[3] It includes a fellowship program for individuals who have strong backgrounds in mathematics and want to become mathematics educators. They go through an extensive five-year program

---

[3]For more information, see http://www.mathforamerica.org/dc [September 2014].

that includes earning a master's degree in mathematics education and teaching in DC secondary schools.

A second program of MfA DC is the MfA Master Teacher Fellowship Program, which began in 2008 and is funded through the MfA headquarters in New York. The program supports five master teachers, including presidential awardees, board-certified teachers, and even the 2014 DC Teacher of the Year, to form a cohort of teachers who can become leaders within their schools and to reach out to teachers outside of their schools on a regional and national level. Master teachers also serve as mentors to the fellows, who in turn are encouraged to become master teachers.

Master teachers have their own set of goals for the level at which they want to work, Jones said, and some are more policy oriented. The program is young and still small, so it has not yet been formally evaluated, but annual evaluations are done as part of the program's support from the National Science Foundation.

## PRESIDENTIAL AWARDS FOR EXCELLENCE IN MATHEMATICS AND SCIENCE TEACHING

The Presidential Awards for Excellence in Mathematics and Science Teaching (PAEMST) Program was enacted by Congress in 1983 to recognize excellent mathematics and science teaching.[4] "It is considered to be the highest recognition a math or science teacher can receive in United States," said the National Science Foundation's Nafeesa Owens, who directs the program. The President has the opportunity to recognize 108 awardees, two from each of the 50 states and from four jurisdictions— Puerto Rico, the District of Columbia, the U.S. Department of Defense education activity schools, and four territories as a group. NSF administers the program on behalf of the Office of Science and Technology Policy in the Executive Office of the President.

Awardees are chosen through a rigorous application process. They complete a narrative that addresses five dimensions of outstanding teaching—content knowledge, teaching strategy, assessment and self-reflection of teaching practice, professional development undertaken, and leadership. Their narrative is accompanied by a video of their teaching as well as recommendations from colleagues, information about the student population that they are reaching, and supplementary information about their teaching. Applications are reviewed at the state level and then again at the national level.

Awardees are recognized through a number of events at the White House and in Washington, DC. They meet with the leadership of federal

---

[4]For more information, see https://www.paemst.org [September 2014].

agencies and Congress and, at the U.S. Department of Education, talk with administrators about activities being implemented in their schools and districts. They also receive $10,000 and a certificate signed by the President.

The program has recognized more than 4,300 awardees. When they return to their schools, said Owens, they serve as "certifiable leaders." They are on committees, help with curriculum change, and participate in endeavors from the school level to the state level. As part of a leadership community, they connect with other leaders and learn about other fellowship programs, including other opportunities to serve in a leadership capacity. "One of the strongest points that the award affords them is the opportunity to meet other excellent teachers and join a community of alumni who are very active within their schools and throughout the nation," said Owens.[5]

One challenge that NSF recognizes is that no formal program currently exists for alumni, said Owens. As a result, the opportunities to which they are exposed remain relatively informal. However, NSF has a working group that is investigating how to formalize the leadership opportunities provided not only to PAEMST awardees but also to other teacher leaders.[6]

## THE NATIONAL ACADEMIES TEACHER ADVISORY COUNCIL

The National Academies' Teacher Advisory Council (TAC) was the direct product of Bruce Alberts' interest in STEM education, said Steve Long, who was chair of the TAC at the time of the convocation.[7] It was established in 2002, when Alberts was president of the National Academy of Sciences, to provide teachers with a regular voice in education policy making. Its mission is "to increase the usefulness, relevance, and communication of research to educational practice; help the research community develop new research that is informed by practice; provide advice about how other National Academies programs, initiatives, and recommendations can be most effectively implemented in schools; and offer guidance about how the National Academies can best communicate with the teaching community in the United States."

The membership of the TAC is diverse, representing all levels of K-12 education and a variety of disciplines. In each grade band, it includes

---

[5]For an analysis of this program from 1994–2006, see SRI International (2009).

[6]An early analysis of the work of three organizations that allow for networking of Presidential Awardee alumni/ae (Council of Presidential Awardees in Mathematics, Association of Presidential Awardees in Science Teaching, and the Society of Elementary Presidential Awardees) is available in Education Development Center, Inc. (1999).

[7]For more information, see http://nas.edu/tac [September 2014].

at least one national board-certified teacher, and others are recipients of Presidential awards or other recognitions. All teachers on the TAC have to spend at least half their time in the classroom, which "gives a lot of weight to what's going on," said Long. "The Teacher Advisory Council brings that wisdom of practice from classroom teacher leaders who are there." The TAC compensates school districts for the cost of substitutes to allow its members to participate in Council meetings and other events.

TAC has collaborated with a wide variety of organizations both within the National Academies and outside. It also works closely with the Einstein fellows. "We refer to each other as cousins because we have so much in common," said Long.

Council members have set up networks among themselves, and the TAC model has been adopted or adapted by state-level teacher advisory councils in California[8] and Michigan.[9] Some TAC members are trying to help establish state-level teacher advisory councils in their home states. The Council has organized or has been associated with a number of workshops and other activities related to teachers and teaching within the National Academies (National Research Council, 2006, 2007, 2014, in preparation;[10] National Research Council and National Academy of Engineering, 2012).

Long noted some issues that the Council as a whole and individual members face. For example, because they are all classroom teachers, when some TAC members leave Washington and return to the day-to-day realities of teaching, it can be "hard to carry the excitement and vision we get in these meetings back into our daily lives," said Long. The group also lacks visibility, he said, which can hinder its ability to work with others.

## THE FULBRIGHT DISTINGUISHED AWARDS
## IN TEACHING PROGRAM

The Fulbright Distinguished Awards in Teaching Program[11] is administered by the International Institute of Education (IIE) under a corporative agreement with the U.S. Department of State's Bureau of Educational and Cultural Affairs. The program is just six years old, said Holly Emert, who leads the program at IIE, but it has been growing quickly. It is open

---

[8]Additional information is available at http://www.ccst.us/ccstinfo/caltac.php [September 2014].

[9]Additional information is available at http://michiganeducators.org [September 2014].

[10]Committee on Strengthening Science Education Through a Teacher Learning Continuum, in association with the NRC Board on Science Education. Available: http://sites.nationalacademies.org/DBASSE/BOSE/CurrentProjects/DBASSE_072083 [September 2014].

[11]Additional information is available at http://www.iie.org/Programs/Fulbright-Awards-In-Teaching [September 2014].

not only to K-12 teachers, but also to guidance counselors, media special-ists, gifted and talented coordinators, special education coordinators, and other K-12 educators. "It's quite rare to find programs that aren't just for classroom teachers," said Emert.

The program funds U.S. educators to go abroad for three to six months, along with teachers from other countries to come to the United States. U.S. educators take one to two courses at a host university, typically in pedagogy or in the study of educational systems. They also visit local schools to observe classes, guest-lecture, and gather project data, lead-ing to a capstone project of direct relevance to their teaching. Emert also listed examples of a broad array of STEM-related projects supported by the program in 2013–2014 that involved visits by teachers from the United States to Finland, Israel, Mexico, and South Africa, along with teachers from Singapore and India who undertook projects in the United States. The other participating countries in 2013–2014 included Chile, Morocco, New Zealand, Palestinian Territories (U.S. teachers only), South Korea (U.S. teachers only), and the United Kingdom (U.S. teachers only).

As benefits to teachers, Emert listed the following:

- Cross-cultural understanding and skills
- Professional development
- Observation and study of international best practices in education
- Sharing of professional expertise with host country teachers and students
- Development of leadership skills
- Development of educator and research networks.

As benefits to schools, she listed the following:

- Integration of global perspectives and methods into classroom practice
- Increased global awareness of students and staff
- Increased ability to work with culturally diverse populations
- Development of cross-national school partnerships.

The program supports 20 to 40 U.S. teachers each year, so it is highly competitive, Emert said. "It gives us the luxury of choosing the most well-developed applications, people who can take what they learn abroad and apply it when they come back," she said. The program encourages teachers to make what they have learned sustainable, "not just to apply it to their classroom but to go farther." It requires that applicants have completed at least five years in the classroom and have gone beyond their usual classroom duties. That may mean building curricula, giving national or state presentations, serving as department chair, or instituting a reform that goes beyond the classroom experience.

Emert concluded by mentioning some of the challenges for the program. It provides no funding for long-term substitutes, so participants have to take a leave of absence with or, far more often, without pay. Also, participants have to leave the country, which is not possible for everyone. Connecting alumni of the program is another challenge, though the State Department does support an alumni group for Fulbright scholars in general.[12] Finally, participants have expressed a desire for mentors to help them bring what they have learned back to the U.S. education system.

## OTHER LEADERSHIP OPPORTUNITIES

Convocation participants mentioned several other professional development options that can build leadership skills. For example, Cindy Hasselbring cited scientific research as a way to develop needed skills and knowledge. Similarly, Ida Chow, Society for Developmental Biology, noted that many professional societies offer opportunities for teachers to engage in professional development activities.

Donald McKinney, Philadelphia Education Fund, described the value of teacher leadership in teacher-led networks designed to create communities of practice. Through such networks, teachers can meet on a regular basis to determine what their students need, how to adjust their instruction to meet those needs, and how to use the resources available to them for professional development. "Teachers know the needs [of their students] best, and they therefore should be both the determiners and the leaders of the professional development that is offered in a school district," McKinney said.

Jay Labov pointed to California, where the California Council on Science and Technology has established a highly successful fellowship program for Ph.D. recipients to go to Sacramento and work in various capacities in the state government.[13] A similar fellowship program is being considered for teachers so that they could work on K-12 education issues that are important in the state. Such programs represent, said Labov, an "inexpensive and yet potentially highly effective model for doing things at the state level."

---

[12]Additional information is available at http://fulbright.org/ [September 2014].
[13]Additional information is available at http://ccst.us/news/2014/0124fellows.php [September 2014].

## TAKING ADVANTAGE OF TEACHER LEADERS

A prominent topic in the discussion session was how to increase the receptiveness of other teachers and administrators to the potential that teachers bring back to schools once they have had a leadership experience out of the classroom. Long pointed out that other teachers and administrators can feel threatened by someone returning to the classroom from elsewhere. They may be jealous of a teacher leader, or they may perceive that knowledge is power and that someone else has therefore become more powerful. Alternately, they may hold a teacher who has had such an experience in greater respect. "It is often difficult to judge which reaction you are going to get from which group or which person," Long said.

Emert observed that these kinds of reactions are one reason why the Fulbright program seeks to build strong networks among participants. Educators returning to a school need to know that some of their colleagues will react positively and some negatively. "Work with those you can," she said. "That's how you encourage change in the school."

Gillespie pointed out that the Knowles Science Teaching Foundation Fellowship does not take teachers out of the classroom. Nevertheless, the program has learned to build structures that teachers can replicate in their schools. For example, it has a heavy focus on practitioner enquiry and professional learning communities. Even small changes, such as Friday pizza lunches for teachers to discuss student learning, can have substantial effects, she said.

Peña noted that the Einstein Fellowship Program has instituted an overnight post-fellowship transition retreat where each cohort of fellows discusses how to leverage its experiences in the fellows' home communities. "This two-day retreat has made significant impact on the long-term goals of the fellowship," she said.

## DIVERSITY IN FELLOWSHIP PROGRAMS

Marilyn Suiter, NSF, brought up a critical issue in fostering and supporting STEM teacher leaders: how to ensure diversity among teacher leaders in terms of ethnicity, gender, and disability. Emert responded that because the Fulbright Program is a government program, it has a mandate to find a diverse pool of scholars. It therefore makes a concerted effort to achieve ethnic, gender, geographic, and other forms of diversity. For example, it works with professional organizations representing minority groups that are underrepresented in STEM fields. It also seeks out individuals who work in schools with underserved populations. Similarly, Long noted that the Teacher Advisory Council tries "to make sure that we have as much diversity within the group as possible to represent all the different viewpoints that we can."

# 4

# Professional Development for STEM Teacher Leaders

Science, technology, engineering, and mathematics (STEM) teacher leaders take many different paths to leadership positions. Professional development can both influence these paths and encourage some teachers to become leaders who would not otherwise have taken that route. Three speakers at the convocation looked at the intersection of professional development and STEM teacher leadership and described ways to strengthen the connection.

## SOLVING THE PUZZLES OF PROFESSIONAL DEVELOPMENT

Suzanne Wilson, Neag Endowed Professor of Teacher Education in the Department of Curriculum and Instruction at the University of Connecticut, organized her analysis of professional development under what she called three puzzles.

### Three Puzzles of Professional Development

The first puzzle Wilson identified involves the design features of professional development. Researchers have sought to identify the attributes that make professional development effective. Among such attributes, said Wilson, are teachers collaborating and actively engaged in professional development, professional development that is focused on content and student learning, professional development that occurs over time, and professional development that is coherent and aligned with relevant policies and practices.

However, when tested on a large scale, professional development with these attributes is not always effective. "We are close to understanding things about high-quality professional development, but when we try it on a large scale in experimental settings, we don't necessarily get the kind of data that would make us confident that these principles would hold," she stated.

One possibility, Wilson said, is that these are surface features of professional development but not actually the things that matter. For example, devoting lots of time or money to professional development does not mean that it will be meaningful. Perhaps time is instead a proxy for the development of trust, which enables people to form the kinds of relationships that they need to learn. "This is one of the big differences between how schools are organized for beginning teachers and how hospitals are organized for beginning doctors," Wilson noted. "Beginning doctors are not responsible for their mistakes. There is a supervising physician who is responsible for that person's mistakes, because they presume that you're going to make mistakes in learning how to practice. . . . There is a transition that needs to be supported by expertise, and we don't have that kind of organization in the structures of schools."

A second puzzle is that most teacher learning does not happen in formal professional development. Instead, teachers learn in many different contexts, many of which are not designated as professional development. These contexts include research involving networks of teachers, preparation for board certification, curriculum or assessment development, engagement in comprehensive school reform, interaction with the leadership in a school, and everyday experiences in the classroom. To understand these much more widespread forms of professional development, researchers need to synthesize across a much wider field of contexts, said Wilson.

The ubiquity of professional development among teachers contrasts with that of nurses, Wilson noted. Beginning nurses know that they are just beginning and that a vision of expert practice exists. They also know that they are on a learning curve and that what they can do now is different from what they will be able to do 10 years from now. "There is a clear vision of a learning progression, and there are planned opportunities for people to intentionally move along that progression, instead of what we do in teaching, which is pretty much leave it to the individual and throw some things that are available at them in the school district when somebody has the resources," she noted.

Wilson also made the point that it may make more sense to talk about the development of workforce capacity in teaching and not the development of individual people, because teachers are always entering and leaving the profession. "But the institutions they work in have memories, and if we work on the collective, there are things that stay in organizations when an individual leaves," she suggested. In addition, as she pointed out, not everyone who teaches is extraordinary, and not everybody who teaches is a learner. Millions of people are teachers, and some of them will not voluntarily step up and engage in the kind of professional development that makes a difference for students.

The third puzzle involves research and experience. Today, she asserted, "research" is seen as objective and free from bias, whereas "experience" is seen as subjective and biased. But researchers use many different sources of information to draw conclusions, not just data, and they often do things that are not based on data but nevertheless work. "Yes, we need research," said Wilson, "but we also need to figure out how to tap into experience in meaningful and productive ways, because there isn't a field in the world that doesn't use experience as a very important source for making decisions." Also, research itself is values laden, Wilson noted, from the questions asked to the methods and instruments used. Teaching can be informed by science, but it is human improvement work, which is not necessarily a science. An excessive focus on science means losing out on an opportunity to tap into teachers' wisdom.

## Conclusions for Research and Practice

From her analysis of the three puzzles, Wilson drew several conclusions regarding professional development. First, researchers need to learn how to do research that can get a handle on the mechanisms at play, she said. For example, the field of evaluation has known for a long time that there are different ways of approaching teacher evaluations that would produce more learning from the evaluation. But most educator evaluation systems have not tapped into that knowledge. "We have to make sure that we know exactly whether assumptions and values are driving the research that we are depending on," she said.

Second, the work of teaching needs to be systematically enhanced so that people can learn from that work, she said, noting that teacher learning also can be more planned and deliberate rather than accidental and serendipitous, and that teacher leadership can have different goals. As she related, leaders can formulate policy, which requires particular skills and knowledge; they can develop expertise, which then can be disseminated; or they can help disseminate expertise developed by others. "That's a different end, and it requires a different skill set," she said. "Learning how to help other people learn is not the same thing as knowing how to teach 10th graders biology. You're dealing with adults who are very different than the children whom you work with, so you need more and different skills."

Third, experience needs to re-enter the dialogue, Wilson stated, even though teachers also need to learn to be critical of experience. In China, for example, teachers have time to study lessons and teaching, but they also have to write about their teaching and present their teaching publicly in front of other teachers for critique.[1] "Yes, teachers are given time, but structures are in place that make people more confident that the time they are using to learn from their experiences is well spent, and we don't have that faith right now in American schools," Wilson said.

The identification of instructional strategies to build leadership remains "one of the big gaps in our knowledge as a field," Wilson concluded. Today, teachers mostly have to figure out how to lead on their own. But people have worked hard on trying to identify some of the pedagogies that can help create leaders, such as coaching, team teaching, induction programs, or what is called motivational interviewing, where the questions asked are designed to encourage particular behaviors. She suggested that a particularly intriguing option would be to map out progressions of teacher learning, including lessons that beginning and more experienced teachers need, so they do not become discouraged and leave the profession.

## PUTTING RESEARCH RESULTS TO USE

In agriculture, medicine, and transportation, new research results can have an immediate effect on practice, noted Arthur Eisenkraft, distinguished professor of science education, professor of physics, and director of the Center of Science and Math in Context at the University of Massachusetts, Boston. However, he said, that is not necessarily the case in education. For one thing, many teachers do not respect research reports, said Eisenkraft, "because they think the reports are written by people who have no idea what goes on in the classroom—and they are often right."

Institutions of higher education have a tendency to lack respect for the wisdom of experience. As an example of the ways in which this wisdom can be incorporated into research, Eisenkraft briefly described a research project that takes advantage of recent changes to Advanced Placement (AP) courses and exams.[2] Teachers have strong incentives to change their

---

[1]For additional background about teachers' work in China, see National Research Council (2010).

[2]For additional information about these changes, see College Board (2011a, 2011b, 2012). This restructuring of AP science courses is based on recommendations in National Research Council (2002).

teaching to accommodate changes in the courses, since their students go on to take AP exams that provide a measure of teaching effectiveness. The research is looking at the choices that teachers made for professional development to accommodate the changes. "Did they go to workshops, did they read things on the Internet, did they come to small groups? . . . Maybe we can find an optimized path: If you've taught for 12 years and you're starting AP bio, take A, B, C, and E, but don't take D, because it won't impact your student scores," he said.

He also mentioned the Wipro Science Education Fellowship, a project to foster teacher leaders in small districts in Massachusetts, New York, and New Jersey.[3] Teachers are interested in leadership, he acknowledged, but not necessarily on the terms imposed by outside initiatives. One reason may be because teachers tend to work as equals, not as hierarchies. He noted, "Maybe teacher leadership is not, 'I know what to do, let me help you.' Maybe it's teachers learning with other teachers."

## DEFINING WHAT WORKS

Janet English, a teacher in the Saddleback Valley Unified School District in California, recounted an experience she had while visiting schools in Finland, from which she had just returned after a six-month Fulbright fellowship. When asked by a university colleague in Finland how a school visit had gone, she replied that it had gone well. Her colleague said, "Stop being so American. Not everything is good. In Finland, we talk about what is working and what is not working—that way we can have a discussion." These are the kinds of discussions that can optimize learning for every child, English said.

Both policy leaders and teachers have complex problems that they need to solve, but they are different kinds of problems. Policy makers have to think about who wins and who loses from a given policy, she noted, while teachers need to think about whether they are engaging the learning of a child. For the education system, therefore, the question is how to solve problems within that system, and "whoever is closest to the problems will have the expertise that's needed to solve the problems best," English said. Teachers are the professionals who make decisions in the classroom. But many other stakeholders influence those decisions, including policy makers, parents, and business people.

In response to a question about how teachers can lead without alienating their colleagues, who may or may not respect the views of a teacher leader, English observed that teacher leaders need to be extremely careful

---

[3] Additional information is available at http://www.umb.edu/cosmic/projects/wipro_science_education_fellowship [September 2014].

when they participate in an activity outside the classroom and come back to their schools to tell other teachers how to do things differently. "Everyone is a teacher leader in their own classroom," English said. "Everyone has value, and everyone comes to the classroom with different styles. I've learned over the years that you can't change a teacher's style. They have ways of doing things." But by concentrating on what is working and what is not working, teachers are less likely to resent outside advice and guidance, she said.

# 5

# Generating Continuous Improvement Cycles Through Research and Development

In her presentation, Suzanne Donovan, executive director of the Strategic Education Research Partnership (SERP), said she received her Ph.D. in the economics of public policy, so she was trained to think about education from an economics perspective. A major emphasis within economics is the idea of incentives, she noted, which in the field of education includes such initiatives as merit pay and competitions among schools. This perspective tends to treat what happens inside the classroom as a black box, Donovan said. "It's not a bad thing to put all that inside a black box," she said. "It makes life a lot easier. It's the great discovery of the invisible hand."

Asking convocation participants to use their cell phones to take an informal survey, Donovan asked them to identify whether they have ever been a K-12 teacher and then to answer the question: "To achieve a high-functioning K-12 education system, what portion of the challenge can be addressed through policies that address incentives and accountability?" Most indicated electronically that they had been teachers, and the large majority of these respondents thought that less than one-quarter of the problem could be solved through incentives and accountability, with a minority choosing between one-quarter and one-half.

Though this sample was small, Donovan acknowledged, the result clearly suggested that treating the classroom as a black box is insufficient. "If we really want to understand how to improve education—and most of you believe this as well—we need to open the black box and see what's inside," she said. This growing realization eventually led

Donovan to switch her interests from economics to the learning sciences, she explained. She worked at the National Research Council (NRC) on the publications, *How People Learn: Brain, Mind, Experience, and School* (National Research Council, 2000) and *How People Learn: Bridging Research and Practice* (National Research Council, 1999), noting this experience made her and her colleagues realize that to affect policy, research results need to be translated in ways that make sense for policy makers. Similarly, she said, teachers need to know what research results "look like" in the classroom to make use of that knowledge. She explained this conclusion led to a third publication, *How Students Learn: History, Mathematics, and Science in the Classroom* (National Research Council, 2005), which incorporated topics from history, mathematics, and science into a representative elementary school-level, middle school-level, and high school-level classroom.

Each chapter required a remarkable amount of work to ensure that the underlying principles were well articulated, said Donovan. But the researchers who helped prepare this publication had a harder time doing so than teachers. "When you get a teacher who is able to think deeply about curriculum, there is a richness that is not there with somebody who hasn't spent years in a classroom," she observed. This kind of translational research, she said, "treats teachers as targets for learning instead of as targets for changing behaviors. Teachers become a more active part of solving the problem of improving instruction."

Yet this translational research also has its limits, she noted. Teachers are trying to balance so many things in the classroom that they can be overwhelmed by the task of incorporating translational research into their teaching. They must simultaneously manage their classrooms, implement engaging activities for their students, and engage in diagnostic teaching. Beginning teachers typically spend much of their time on the first and second of these activities. "For these teachers, translational research isn't going to help," said Donovan. More accomplished teachers can spend more of their time on implementing engaging activities than on classroom management, but even they tend to spend little time engaged in diagnostic teaching. Only with expert teachers, who represent a small fraction of all teachers, is a substantial fraction of their time devoted to diagnostic teaching.

As many others have pointed out, she said, schools are complex systems, and teacher expertise is only one factor in a system where the parts interact with each other in unpredictable ways. In such systems, said Donovan, "it is very difficult to get education change—and impossible to get it from the outside." A variety of obstacles can stand in the way of moving toward best practices, which requires that many people be involved in removing those obstacles.

## MOVING RESEARCH INTO SCHOOLS

This gradual evolution of thinking about teaching at the National Academy of Sciences led to an effort, staffed by Donovan, to do work on teaching practices in a practice setting, not in research universities or the National Academies. The report *Strategic Education Research Partnership* (National Research Council, 2003) argued that a new kind of organization was needed to move the research enterprise into school settings, with problems taken not from theory but from practice. The result was the creation of the independent nonprofit Strategic Education Research Partnership (SERP), which Donovan now heads. Its theory of action "puts practitioners, both teachers and administrators, into a much more central role," she said, explaining it brings together education professionals, researchers, and designers in principled collaborations to work on tools, programs, and practices that can result in achievement gains and knowledge accumulation at scale.

Donovan emphasized the role of designers. They understand problems from a different perspective, she said. They always begin with the user and how a tool, program, or practice is going to influence the user's behavior. "What's in it for them, what's going to make them change their behavior? If the answer is nothing, don't expect that what you hope they're going to do is in fact what they're going to do," she commented.

The process at SERP involves proposing and designing solutions, testing them in practice settings, finding whether they work or not, and making iterative revisions until something does work. Teachers pilot what works, and according to Donovan, "the extent to which the final product works in practice is largely an artifact of how engaged teachers have been in shaping the final product."

Donovan demonstrated several examples of products devised through this process. One of the most popular has been a collection of five-inch by eight-inch cards that list the principles that a lesson is designed to achieve and the vital actions expected of students. The idea was to avoid giving teachers huge binders filled with aspirations. Rather, the cards are designed to give teachers something that they can use to figure out why they need to change. They do not have to read large amounts of material to change their teaching. What they need to know in a particular circumstance is on the card. This approach "treats teachers differently," said Donovan. "It treats them as an essential member of a team and as clients for whom the design team needs to design."

Classrooms taught using this approach can look more chaotic than normal classrooms, but they also can increase student learning, Donovan said. One challenge is to get principals to "change their taste in instruction" so that they are not put off when they walk into a classroom using these materials. The five-by-eight cards have been critical in this respect, because they let principals know why a classroom might look different than they expect.

A deeper problem is the need to shift policy makers away from thinking solely in terms of incentives and accountability as the primary drivers of change, said Donovan. "It's not that [incentives] are irrelevant; it's just that they're not going to solve very much of the problem," she concluded. "The basic problem is building capacity and solving the problems of a complex system. We need to shift to an investment in a very different kind of policy."

## ENGAGING PARENTS

In the discussion during Donovan's presentation, a workshop participant pointed out that having five-by-eight cards not only for teachers and principals, but also for parents would help parents understand new approaches to teaching. Donavan responded that SERP has not worked on that problem yet, but it is an important problem that probably will receive attention in the future.

Rebecca Sansom, a 2013–2014 Einstein fellow, raised the point that whenever teachers try something new, they are not going to do that new thing as well as they did an older thing, even if the newer approach will ultimately be more effective than the older approach. This is an issue of teacher leadership, she said, "because there is something about teacher leaders that makes them more willing to accept that risk and that vulnerability."

Donovan pointed out that the key is to clarify the purpose of change. Teachers need to understand the reason for a change and then be supported to make the change and incorporate it into their practice, she said. One useful option, she said, might be to create schools within a district that are focused on innovation, and then put expert master teachers into those schools. New and more experienced teachers then could work with or in these schools to incorporate new materials into their practices. Such an approach could also help bring about an essential change in teaching, said Donovan, toward a profession that is widely viewed as highly skilled.

# 6

# Concluding Observations

In the final session of the convocation—and periodically at earlier points in the meeting—participants were invited to offer their general reflections on the issues raised by the presenters. Those comments are summarized in this last chapter of the report to review some of the messages expressed by many participants at the convocation.

## THE ROLE OF ADMINISTRATORS

Many of the commenters called attention to the importance of administrators in fostering and enabling STEM (science, technology, engineering, and mathematics) teacher leadership. For example, Cindy Hasselbring, Maryland State Department of Education, pointed out that administrators and district leaders play a key role in identifying teacher leaders, inculcating the skills of these teachers, and giving them a voice.

At the same time, teachers need to help administrators understand that they know what students need and that they need the flexibility to connect with students, said Toby Horn, Carnegie Institution for Science. As Camsie McAdams, U.S. Department of Education, noted, the leadership in a school or district may not have STEM expertise, yet they are making decisions that directly affect STEM education. STEM teacher leaders can counterbalance this lack of expertise. "We need to empower ourselves," she said.

Sophia Gershman, Watchung Hills Regional High School, broadened the discussion by observing that empowering teachers is different than fostering teacher leaders. She also emphasized the importance of including parents in the conversation. "We talked about educating local boards of education and administrators, but we also have to involve parents," she stated.

Similarly, other participants pointed to a wide variety of assets that teachers can use in changing policies that affect teacher leadership. Many organizations exist that are focused on policy, said Steve Robinson, Democracy Prep Charter High School, such as the National Science Teachers Association, the National Council of Teachers of Mathematics, or the new organization Teach Plus, which trains teachers to think about a career path involving policy making.[1] "They know how to do this," said Robinson. "We should think about what those assets are." Another valuable resource, said McAdams, is the Teach to Lead organization, which seeks to catalyze fundamental changes in the culture of schools and teaching so that teachers play a more central role in transforming teaching and learning and in the development of policies that affect their work.[2]

## THE ROLE OF PROFESSIONAL DEVELOPMENT

Participants also focused on the contributions that professional development can make toward creating a robust corps of STEM teacher leaders. As Mike Town, Tesla/STEM High School, said, many teachers who come from science backgrounds had opportunities to remain in or return to science but stayed in education because they had a great professional development experience. In that regard, high-quality professional development is a way to keep good teachers in education and prepare them for leadership positions.

"Good professional development changes a teacher as a person," said Gershman. "What we do to change our skill set immediately affects the classroom. . . . We change ourselves, we change our classroom, we change our students."

Sansom pointed to the importance of teacher preparation as a time for fostering teacher leaders. By creating reflective practitioners, teacher preparation programs can build the skills for leadership and communication with peers and administrators at a school. "There are specific skills that we can learn and that we don't all possess," she said. As another opportunity, Einstein fellow Sheryl Sotelo mentioned the importance of bringing elementary teachers and early childhood educators up to speed on STEM education.

---

[1]For more information about Teach Plus, see http://www.teachplus.org [September 2014].
[2]For more information about Teach to Lead, see http://teachtolead.net [September 2014].

Many levels and kinds of resources need to line up for professional development to work well, said Dorothy Fleisher, W.M. Keck Foundation. For example, she suggested that professional societies and foundations could think about how to involve administrators in professional development, since they are often the ones making decisions regarding the use of new knowledge in their schools.

## THE PROMISE OF STEM TEACHER LEADERSHIP

Finally, many participants pointed to the benefits that STEM teacher leaders can bring to teachers, students, and the education system in general. As Town said, "The more we are influencing decision makers, the better the profession can be."

Town drew a distinction between mentor teachers who work with other teachers and master teachers who work on policy. One possible option would be to identify 435 policy master teachers—one for every congressional district—and fund them from the federal level, with a matching requirement from the state level. These master teachers could work in a congressional office, in a state office, on a legislative committee, on a standing committee for a governor, for industry, for one of the state organizations that are part of the growing STEMx network,[3] or in some other capacity. The result would be statewide and national networks of policy experts in STEM education issues that could have an influence on all levels of the education system. "It's not a perfect solution, but I'm putting it out there as a way of thinking about potential mechanisms that can get political buy-in," Town said.

Robinson referred to the power of collective action among STEM teachers. Thousands of STEM teachers have been recognized at the state and national levels for their accomplishments, yet they have not been organized into a cohesive group, which is "an opportunity that's been missed," he said.

For collective action to be effective, said Heidi Schweingruber, National Research Council, teacher leaders need to have objectives and measurable ways of achieving those objectives. "What collective action could a national network most effectively be involved in? It might be responding to policy issues, it might be trying to set an agenda for policy. Describing those might galvanize more people to get involved because they might see that here is a way that I can make a difference."

Richard Duschl, National Science Foundation, emphasized the importance of better coordination among the federal agencies that nurture and support teacher leaders. Agencies pursue these activities in different ways

---

[3]More information is available at http://www.stemx.us [September 2014].

and often in isolation. "It's a quality control issue I'm alluding to, and the best people to help us with that quality control are the fellows themselves," he said. For example, current fellows can provide suggestions for future fellows, not only within but also across programs.

Nancy Arroyo, Riverside High School, suggested taking on a specific problem that is general enough for everyone to buy into and specific enough to be solvable. "If we can do that, even if it takes five years, the time and money would be well spent," she said. Robinson similarly noted that getting teachers involved in policy is too large to be a discrete and doable task. He asked, "What is the small discrete task that could come out of the convocation as a way of advancing its core objective?"

Juliana Jones, Longfellow Middle School, reminded the group that this is a unique time in education with the advent of the Common Core State Standards and other changes in education research and practice. Effective leadership could help create changes that touch every classroom in America. The result, agreed teacher Janet English, would be to ensure that the needs of students remain at the center of what is done in the classroom and in policy making. Teachers "have to be the voice of what helps optimize learning for each and every child," she said.

# References

Alberts, B.A. (2013). Prioritizing science education. *Science, 340,* 249.

American Chemical Society. (2008). *Chemistry in the National Science Education Standards, Second Edition.* Washington, DC: Author.

American Chemical Society. (2012). *The ACS Guidelines and Recommendations for the Teaching of High School Chemistry.* Washington, DC: Author.

Berry, B. (2011). *Teaching 2030: What We Must Do for Our Students and Our Public Schools— Now and in the Future.* New York: Teachers College Press and Washington, DC: NEA Professional Library.

Bhatt, M.P., and Behrstock, E. (2010). *Managing Educator Talent: Promising Practices and Lessons from Midwestern States.* Naperville, IL: Learning Point Associates.

Bissaker, K., and Heath, J. (2005). Teachers' learning in an innovative school. *International Education Journal, 5*(5), 178–185.

College Board. (2011a). *AP Biology Curriculum Framework 2012–2013.* New York: Author.

College Board. (2011b). *AP Chemistry Curriculum Framework 2013–2014.* New York: Author.

College Board. (2012). *AP Physics 1: Algebra-based and AP Physics 2: Algebra-based Curriculum Framework 2014–2015.* New York: Author.

Education Development Center, Inc. (1999). *National Alliance of Presidential Awardee Associations. Conference Proceedings.* October 1–4, 1998, Baltimore, MD. Newton, MA: Author.

Kimmelman, P.J. (2010). *The School Leadership Triangle: From Compliance to Innovation.* Thousand Oaks, CA: Corwin Press.

MetLife Foundation. (2012). *The MetLife Survey of the American Teacher: Challenges for School Leadership, A Survey of Teachers and Principals.* New York: Author. Available: http://files.eric.ed.gov/fulltext/ED542202.pdf [October 2014].

National Research Council. (1990). *Fulfilling the Promise: Biology Education in the Nation's Schools.* Washington, DC: National Academy Press.

National Research Council. (1999). *How People Learn: Bridging Research and Practice.* M.S. Donovan, J.D. Bransford, and J.W. Pellegrino (Eds.). Committee on Learning Research and Educational Practice. Commission on Behavioral and Social Sciences and Education. Washington, DC: National Academy Press.

National Research Council. (2000). *How People Learn: Brain, Mind, Experience, and School, Expanded Edition.* Committee on Developments in the Science of Learning. J.D. Bransford, A.L. Brown, and R.R.Cocking (Eds.) with additional material from the Committee on Learning Research and Educational Practice, M.S. Donovan, J.D. Bransford, and J.W. Pellegrino (Eds.), Committee on Learning Research and Educational Practice, Commission on Behavioral and Social Sciences and Education. Washington, DC: National Academy Press.

National Research Council. (2002). *Learning and Understanding: Improving Advanced Study of Mathematics and Science in U.S. High Schools.* Committee on Programs for Advanced Study of Mathematics and Science in American High Schools, J.P. Gollub, M.W. Bertenthal, J.B. Labov, and P.C. Curtis (Eds.), Center for Education, Division of Behavioral and Social Sciences and Education. Washington, DC: National Academy Press.

National Research Council. (2003). *Strategic Education Research Partnership.* Committee on a Strategic Education Research Partnership. M.S. Donovan, A.K. Wigdor, and C.E. Snow (Eds.), Division of Behavioral and Social Sciences and Education. Washington, DC: The National Academies Press.

National Research Council. (2005). *How Students Learn: History, Mathematics, and Science in the Classroom.* Committee on How People Learn, A Targeted Report for Teachers, M.S. Donovan and J.D. Bransford (Eds.), Division of Behavioral and Social Sciences and Education. Washington, DC: The National Academies Press.

National Research Council. (2006). *Linking Mandatory Professional Development with High-Quality Teaching and Learning: Proceedings and Transcripts.* Teacher Advisory Council, Division of Behavioral and Social Sciences and Education. Washington, DC: The National Academies Press.

National Research Council. (2007). *Enhancing Professional Development for Teachers: Potential Uses of Information Technology, Report of a Workshop.* Committee on Enhancing Professional Development for Teachers, Teacher Advisory Council, Center for Education, Division of Behavioral and Social Sciences and Education. Washington, DC: The National Academies Press.

National Research Council. (2010). *The Teacher Development Continuum in the United States and China: Summary of a Workshop.* A. Ferreras and S. Olson, Rapporteurs. A.E. Sztein (Ed.). U.S. National Commission on Mathematics Instruction, Board on International Scientific Organizations, Policy and Global Affairs. Washington, DC: The National Academies Press.

National Research Council. (2012). *A Framework for K-12 Science Education: Practices, Crosscutting Concepts, and Core Ideas.* Committee on a Conceptual Framework for New K-12 Science Education Standards. Board on Science Education, Division of Behavioral and Social Sciences and Education. Washington, DC: The National Academies Press.

National Research Council. (2014). *STEM Learning Is Everywhere: Summary of a Convocation on Building Learning Systems.* S. Olson and J. Labov, Rapporteurs. Planning Committee on STEM Learning Is Everywhere: Engaging Schools and Empowering Teachers to Integrate Formal, Informal, and After-School Education to Enhance Teaching and Learning In Grades K-8, Teacher Advisory Council, Division of Behavioral and Social Sciences and Education. Washington, DC: The National Academies Press.

National Research Council and National Academy of Engineering. (2012). *Community Colleges in the Evolving STEM Education Landscape: Summary of a Summit.* S. Olson and J.B. Labov, Rapporteurs. Planning Committee on Evolving Relationships and Dynamics Between Two- and Four-Year Colleges and Universities. Board on Higher Education and Workforce, Division on Policy and Global Affairs. Board on Life Sciences, Division on Earth and Life Studies. Board on Science Education, Teacher Advisory Council, Division of Behavioral and Social Sciences and Education. Engineering Education Program Office, National Academy of Engineering. Washington, DC: The National Academies Press.

Organisation for Economic Co-operation and Development. (2011). *Building a High Quality Teaching Profession: Lessons from Around the World*. Background Report for the International Summit on the Teaching Profession. Paris: Author.

Pennington, K. (2013). *New Organizations, New Voices: The Landscape of Today's Teachers Shaping Policy*. Washington, DC: Center for American Progress.

Rasberry, M.A., and Mahajan, G. (2008). *From Isolation to Collaboration: Promoting Teacher Leadership through PLCs*. Hillsborough, NC: Center for Teaching Quality.

SRI International. (2009). *Synthesizing Evaluation Results for NSF's Presidential Awards for Excellence in Mathematics & Science Teaching (PAEMST) Program*. Available: http://csted.sri.com/projects/synthesizing-evaluation-results-nsfs-presidential-awards-excellence-mathematics-science-tea [September 2014]

Teacher Leadership Exploratory Consortium. (2011). *Teacher Leader Model Standards*. Carrboro, NC: Center for Teaching Quality.

# Appendix A

# Convocation Agenda

**June 5-6, 2014**
**Lecture Room**
**National Academy of Sciences Building**
**2101 Constitution Avenue, NW**
**Washington, DC**

*Thursday, June 5*

**7:30 am**   **Registration and Networking**
Registration table and full breakfast available outside of
Lecture Room

**8:00 am**   **Welcome, Introductions, Overview, and Learning Goals
for the Convocation**
**Summary of Participant Responses to Registration
Questions**
• Mike Town, representing the planning committee
• Steve Long, representing the Teacher Advisory Council
• Karen King, representing the National Science Foundation
• Jay Labov, convocation overview and logistics

**8:15 am**      **Keynote Presentation and Discussion**
*Empowering Our Best Teachers: A Grand Challenge in Education*
- Bruce Alberts, former President of the National Academy of Sciences and former Editor-in-Chief of *Science*

**8:45 am**      **Session of Panelists and Discussion**
*What are the opportunities and challenges of giving a voice to teachers around education policy and other forms of decision making at the national, state, and local levels?*

A panel will provide perspectives about how teachers are involved with these issues, the benefits that teacher voices have brought to the table, and any challenges faced. Convocation participants will be able to ask general questions and additional discussion is scheduled for 12:00 today following additional perspective from two plenary speakers.

PANELISTS
- Diane Briars, President, National Council of Teachers of Mathematics
- Francis Eberle, National Association of State Boards of Education
- Cindy Hasselbring, Maryland State Department of Education
- Steven Robinson, Democracy Prep Charter High School (formerly with the U.S. Department of Education and Executive Office of the President)
- Terri Taylor, American Chemical Society

**9:30 am**      **Plenary Session and Discussion of "Grand Challenges in Education" Paper**
*Teacherpreneurs: Leaders For Tomorrow's Schools*
- Barnett Berry, Center for Teaching Quality

**10:30 am**    **Break and Networking**

**11:00 am**    **Plenary Session and Discussion of "Grand Challenges in Education" Paper**
*Effective STEM Teacher Preparation, Induction, and Professional Development*
- Suzanne Wilson, University of Connecticut, Storrs

**12:00 pm**     **Reactions to Plenary Sessions and Moderated Discussion**
- Janet English, Teacher, El Toro High School (CA) and Member of the Organizing Committee
- School Administrator (invited)
- Arthur Eisenkraft, University of Massachusetts, Boston
- Moderator: Margo Murphy, Camden Hills High School (TAC)

**12:30 pm**     **Lunch and Networking**.
Lunch available outside the Lecture Room
Sign up for afternoon breakout sessions.

**1:30 pm**     **Overview of Additional Current Models of Engaging Teachers' Voices in Education Issues**
- Einstein Distinguished Educator Fellows, Triangle Coalition, Anthonette Peña
- Knowles Science Teaching Foundation Fellowships, Nicole Gillespie
- Math for America, Marlena Jones
- Presidential Awards for Excellence in Math and Science Teaching, National Science Foundation, Nafeesa Owens
- Teacher Advisory Council, Steven Long (TAC Chair)
- Fulbright Distinguished Awards in Teaching Program, Institute of International Education, Holly Emert
- Moderator: Jay Labov

**2:40 pm**     **Break and Move to Breakout Session with Presenter(s) of Your Choice**

**2:45 pm**     **Breakout Sessions with the Presenters**
*Continue discussions with one of the presenters of your choice. The goal of these facilitated discussions is to explore the implications of these papers for policy, implementation, and teacher practice.*
- Professional development to engage teachers in leadership activities (Suzanne Wilson): Room 120 (Heidi Schweingruber, NRC, Facilitator)
- Teacherpreneurs (Barnett Berry): Members' Room (Toby Horn, Facilitator)
- Current models of engaging teachers in leadership activities (Panelists): Lecture Room (Janet English, Facilitator)

**3:55 pm**     **Move to Second Breakout Session of Your Choice**

**4:00 pm**     **Breakout Sessions with the Presenters**
*Continue discussions with one of the presenters of
your choice. The goal of these facilitated discussions is
to explore the implications of these papers for policy,
implementation, and teacher practice.*

- Professional development to engage teachers in
  leadership activities
  (Suzanne Wilson): Room 120 (Mike Town, Facilitator)
- Teacherpreneurs (Barnett Berry):
  Members' Room (Toby Horn, Facilitator)
- Current models of engaging teachers in leadership
  activities (Panelists):
  Lecture Room (Janet English, Facilitator)

**5:00 pm**     **Break and Reassemble in Lecture Room**

**5:15 pm**     **Lessons Learned**
*Facilitators who were assigned to work in each breakout room
will provide an overview of the most important points, insights,
or avenues for future research that emerged from each of their
sessions. Followed by general group questions and discussion.*

**6:00 pm**     **Adjourn and Dinner on Your Own**
List of suggestions for restaurants is at the registration table
outside the Lecture Room.

*Friday, June 6*

**7:30 am**     **Networking**
Full breakfast available outside the lecture room

**8:00 am**     **Recap of Day 1 by Members of the Organizing Committee**
- Janet English, Cindy Hasselbring, Toby Horn,
  Steve Robinson, Mike Town

**8:30 am**　　　**3-2-1 Exercise**
　　　　　　　*Each participant will submit on an index card for later*
　　　　　　　*discussion:*
- Three things that you have learned from the
  convocation thus far.
- Two things about which you still have questions.
- One thing that you will take away from this event that
  can be applied in your home setting.

**9:00 am**　　　**Plenary Session and Discussion**
　　　　　　　*Generating Continuous Improvement Cycles Through*
　　　　　　　*Research And Development in School Systems*
- Suzanne Donovan, Strategic Education Research
  Program

The goal of this longer session will be to explore the
implications of the vision for future research for policy,
implementation, and teacher practice.
- Moderator: Juliana Jones (TAC)

**10:45 am**　　**Break and Networking**
Move to breakout sessions at 11:05 am.

**11:15 am**　　**Breakout Sessions**
*In these sessions participants who represent a sector*
*of the education system will meet together and with*
*representatives from their professional societies to*
*develop and agenda for action. Each group will appoint*
*a rapporteur and prepare not more than two PowerPoint*
*slides that provide the following information:*
- One or two actions that people in your sector can
  implement *at the beginning* of the new academic year.
- One or two actions that people in your sector can
  accomplish *by the end* of the next academic year.
- At least one other sector that can help you achieve your
  goals by working collaboratively.

- Teachers (Room 120)
- Teacher educators (Room 125)
- School administrators (Room 120)
- Education researchers (Lecture Room)
- Funding agencies and organizations (Members' Room)
- Education policy makers (Lecture Room)

Lunch will be available beginning at 11:45 AM.
Return to the Lecture Room beginning at 12:25.

**12:30 pm**    **Each breakout session facilitator from Day 2 will provide
a summary of that group's action plan, answer questions,
and engage participants in discussion.**

**1:30 pm**     **Summary Comments by Members of the Organizing
Committee and Representative from the National Science
Foundation followed by general discussion.**

**2:00 pm**     **Convocation Ends**

# Appendix B

# Convocation Participants

† Presenter
* Member, Convocation
    Organizing Committee
** Member, National Academies
    Teacher Advisory Council
*** NRC Staff

## IN-PERSON PARTICIPANTS

Bruce Alberts**,†
Chancellor's Leadership Chair in
    Biochemistry and Biophysics
    for Science and Education
University of California, San
    Francisco

Nancy Arroyo**
Teacher and Member of the
    National Academies Teacher
    Advisory Council
Riverside High School, TX

Gail Atley
High School Science Teacher
Inglewood Unified School District,
    CA

Ophelia Barizo
Einstein Fellow 2013–2014 (NSF)
Vice Principal for Advancement/
    STEM Coordinator
Highland View Academy, MD

Sarah Bax**
Mathematics Teacher
DC Public Schools, DC

Kathleen Bergin
Associate Lead Program Director
National Science Foundation

Barbara Berns
Senior Project Manager
Education Development Center,
    Inc., MA

Barnett Berry†
CEO and Founder
Center for Teaching Quality, NC

Elaine Blomeyer
Einstein Fellow 2013–2014 (NSF)
Computer Science and
    Mathematics Instructor
Greater Los Angeles Area, CA

Diane Briars†
President, National Council of
    Teachers of Mathematics, PA

Barbara Buckner
Einstein Fellow 2013–2014 (NSF)
Mathematics, Chemistry, and
    Physics Teacher
Bradley Central High School, TN

David Campbell
Program Director
National Science Foundation, VA

Betty Carvellas†
Teacher Leader, National
    Academies Teacher Advisory
    Council, VT

Ida Chow
Executive Officer
Society for Developmental
    Biology, MD

Suzanne Donovan†
Executive Director
Strategic Education Research
    Partnership, DC

Richard Duschl
Senior Advisor
National Science Foundation, VA

Francis Eberle†
Acting Deputy Executive Director
National Association of State
    Boards of Education, VA

Arthur Eisenkraft†
Distinguished Professor of Science
    Education
University of Massachusetts
    Boston, MA

Holly Emert†
Assistant Director, Global Teacher
    Programs Division
Institute of International Education,
    DC

Janet English*,†
Science Teacher
El Toro High School, CA

David Evans
Executive Director
National Science Teachers
    Association, VA

Evelina Felicite-Maurice
Professional Development Trainer
NASA/SSAI, MD

Dorothy Fleisher
Program Director
W.M Keck Foundation, CA

Catherine Fry
Project Manager
Association of American Colleges
    and Universities, DC

Sophia Gershman
Teacher
Watchung Hills Regional High
    School, NJ

Winnie Gilbert**
Teacher
Los Altos High School, CA

Nicole Gillespie†
Executive Director
Knowles Science Teaching
    Foundation, NJ

Bruce Grant
Professor of Biology and
    Environmental Science
Widener University, PA

Natalie Harr
Einstein Fellow (NSF)
Early Childhood Educator (PreK-3)
Crestwood Primary School, OH

Cindy Hasselbring*,†
Special Assistant to the State
    Superintendent
Maryland State Department of
    Education

Heidi Haugen
Agricultural Science Education
    and Chair, California Teacher
    Advisory Council
Florin High School, CA

Robert Hilborn
Associate Executive Officer
American Association of Physics
    Teachers, MD

Toby Horn*†
Co-Director, CASE
Carnegie Institution for Science, DC

Arundhati Jayarao
Chief Operating Officer,
BLUECUBE Aerospace, VA

Juliana Jones**
Algebra Teacher
Longfellow Middle School, CA

Marlena Jones
Master Teacher Program
    Coordinator
Math for America DC

Mary Ann Kasper***
Senior Program Assistant
National Research Council, DC

Susan Kelly
Graduate Student and STEM
    Education Specialist
University of Maryland, Eastern
    Shore

Diane Jass Ketelhut
Associate Professor of Science,
    Technology, and Math
    Education
University of Maryland, College
    Park, MD

Karen King†
Program Director
National Science Foundation, VA

Stephanie Knight
Associate Dean for Undergraduate
    and Graduate Studies
Pennsylvania State University, PA

Andy Kotko
Teacher
Mather Heights Elementary
    School, CA and California
    Teacher Advisory Council

Ken Krehbiel
Associate Executive Director for
    Communications
National Council of Teachers of
    Mathematics, VA

Jay Labov***
Senior Advisor for Education and
    Communication and Staff
    Director, National Academies
    Teacher Advisory Council
Washington, DC

Matthew Lammers***
Program Coordinator
National Research Council, DC

Steve Long**,†
Chemistry Teacher
Rogers High School, AR

Jennie Lyons
Einstein Fellow
National Science Foundation, VA

Camsie McAdams
Senior STEM Advisor
U.S. Department of Education, DC

Melissa McCartney
Associate Editor, *Science*
American Association for the
    Advancement of Science, DC

Catherine McCulloch
Project Director
Education Development Center,
    Inc., MA

Donald McKinney
Program Coordinator
Philadelphia Education Fund, PA

Zovig Minassian
Einstein Fellow 2013–2014 (U.S.
    Department of Energy)
Biology Teacher
Glendale Unified School District,
    CA

Margo Murphy**,†
Science Teacher
Camden Hills Regional High
    School, ME

Barbara Olds
Senior Advisor, Directorate
    on Education and Human
    Resources
National Science Foundation, VA

Steve Olson
Consultant Writer
Seattle, WA

Nafeesa Owens†
Program Director
National Science Foundation, VA

Anthonette Peña†
Program Director for the Einstein
    Fellows
Triangle Coalition for STEM
    Education, VA

Angela Phillips Diaz
Washington Liaison for Education,
    Science, and Technology
California Council on Science and
    Technology, DC

Stephen Portz
Einstein Fellow (NSF)
Space Coast Junior/Senior High
    School, FL

Joan Prival
Program Director
National Science Foundation, VA

Kendra Renae Pullen**
Teacher
Riverside Elementary School, LA

Lynn Foshee Reed
Einstein Fellow (NSF)
Mathematics Instructor
Maggie L. Walker Governor's
    School, VA

Steve Robinson*, **
Science Teacher
Democracy Prep Charter High
    School, NY

Rebecca Sansom
Einstein Fellow 2013–2014 (NSF)
Assistant Professor of Chemistry
Brigham Young University, UT

Heidi Schweingruber***
Director, Board on Science Education
National Research Council, DC

Jennifer Sinsel**
Gifted and Talented Teacher
Wichita Public Schools, KS

Gerald Solomon
Executive Director
Samueli Foundation, CA

Sheryl Sotelo
Einstein Fellow (NSF)
Elementary Teacher

Marilyn Suiter
Program Director
National Science Foundation, VA

Terri Taylor†
Assistant Director, K-12 Education
American Chemical Society, DC

David Thesenga
Einstein Fellow (NSF)
Science and Engineering Instructor
Timberline School, CO

Jim Town
Einstein Fellow 2013–2014 (NSF)
Mathematics Specialist
Alameda County Office of
    Education, CA

Mike Town*,**,†
Teacher
Tesla STEM High School, WA

Elizabeth VanderPutten
Program Director
National Science Foundation, VA

Jo Anne Vasquez
Vice President for Educational
    Practice
Helios Education Foundation, AZ

Claudia Walker**
Teacher
Murphey Traditional Academy, NC

Suzanne Wilson†
Professor of Education
University of Connecticut, Storrs

David Yarmchuk
Curriculum Specialist for Math
    and Science
Cesar Chavez Public Charter
    School, DC

## WEBCAST REGISTRANTS

Barbara Cagni
Teacher
Hobe Sound Elementary School
   (FL)

Raghda Daftedar
Doctoral Candidate
Teachers College, Columbia
   University, NY

Grace Doramus
Chief of Staff
100Kin10, NY

Kris Gutierrez
Professor
University of Colorado, Boulder

Henry Heikkinen
Professor Emeritus
University of Northern Colorado

Diana Kasbaum
Mathematics Education Consultant
Association of State Supervisors of
   Mathematics, WI

Michael Lach
Director of STEM and Strategic
   Initiatives
University of Chicago, IL

Felicia Mensah
Associate Professor and Program
   Coordinator
Teachers College
Columbia University, NY

Harold Pratt
President
Science Curriculum, Inc., CO

Kellie Taylor
K-5 Engineering Teacher
Galileo STEM Academy, ID

Emily Vercoe
Director, Next Steps
Earth Force, GA

Constance Williams
Educator and Department Chair
Fairfield-Suisun Unified School
   District, CA

Elizabeth Young
Teacher
Snapfinger Elementary School, GA

# Appendix C

# Biographical Sketches of Presenters, Panelists, Moderators, and Planning Committee Members

**Bruce Alberts** (Presenter) has served as editor-in-chief of *Science* and as one of President Obama's first three U.S. Science Envoys. He holds the Chancellor's Leadership Chair in Biochemistry and Biophysics for Science and Education at the University of California, San Francisco, to which he returned after serving two six-year terms as the president of the National Academy of Sciences (NAS).

**Barnett Berry** (Presenter) is founder, partner, and chief executive officer at the Center for Teaching Quality, a national nonprofit organization based in Carrboro, North Carolina. He is a former classroom teacher, think tank analyst, state education agency executive, and university professor.

**Diane J. Briars** (Panelist) is president of the National Council of Teachers of Mathematics. She was a mathematics education consultant and a senior developer and research associate for the National Science Foundation-funded Intensified Algebra Project and mathematics director for the Pittsburgh Public Schools. She began her career as a secondary mathematics teacher.

**Betty Carvellas** (Moderator) retired in 2007 after teaching science for 39 years at the middle and high school levels. She was a founding member of the National Academies Teacher Advisory Council (TAC) and currently serves as the Teacher Leader for TAC. In 2008, she was designated a lifetime National Associate of the National Research Council of the National Academies.

**Suzanne Donovan** (Presenter) is executive director of the Strategic Education Research Partnership (SERP), an organization that conducts programs of problem-solving research, design, and development in partnership with school districts and university researchers. Prior to founding SERP, Donovan was a study director at the National Research Council.

**Francis Eberle** (Panelist) is acting deputy executive director for the National Association of State Boards of Education (NASBE). He also acts as director of NASBE's Next Generation Science Standards project. He is an adjunct professor at George Mason University and was a middle school science teacher for 15 years.

**Arthur Eisenkraft** (Panelist) is the distinguished professor of science education, professor of physics, and director of the Center of Science and Math in Context at the University of Massachusetts, Boston. For 25 years, he taught high school physics and was a grade 6-12 science coordinator. He served on the content committee and helped write the *National Science Education Standards* and has served on other National Research Council committees.

**Holly Emert** (Panelist) is an assistant director in the Global Teacher Programs Division at the Institute of International Education in Washington, DC. She manages the Fulbright Distinguished Awards in Teaching Program. She is a former teacher who has taught in China, France, and the United States.

**Janet English** (Committee Member, Panelist) has taught middle and high school science in the Saddleback Valley Unified School District of California since 1992. In 2013, she received the Fulbright Distinguished Award in Teaching and traveled for six months across Finland. She is an associate member of the National Academies Teacher Advisory Council and was the founding vice chair of the California Teacher Advisory Council.

**Nicole Gillespie** (Panelist) leads the Knowles Science Teaching Foundation. Her experience in science, technology, engineering, and mathematics education includes teaching at Menachem Mendel Seattle Cheder High School in Washington and in the Upward Bound Program at Napa Valley College, as well as working with the Seattle Public Schools. She has taught science and education courses at the University of Washington, University of California, Berkeley, and University of Pennsylvania.

**Cindy Hasselbring** (Committee Member, Panelist) serves as special assistant to the state superintendent for special projects at the Maryland State

Department of Education. In this role, she coordinates efforts in science, technology, engineering, and mathematics education and digital learning. She recently completed her second year as an Albert Einstein distinguished educator fellow. Hasselbring taught mathematics for 16 years at Milan High School in Milan, Michigan.

**Toby Horn** (Committee Member) is co-director of the Carnegie Academy for Science Education at the Carnegie Institution for Science in Washington DC. She developed and taught classes at Thomas Jefferson High School for Science and Technology, in Fairfax County, Virginia, and was outreach coordinator at the Fralin Biotechnology Center at Virginia Tech. She currently serves on the National Visiting Committee for the National Science Foundation-funded Bio-Link Center of Excellence.

**Juliana Jones** (Moderator) teaches Algebra I at Longfellow Middle School in the Berkeley Unified School District in California. Jones has served as a teacher leader for the Bay Area Math Project and is an original member of the California Teacher Advisory Council. She was the only teacher to serve on the California Council on Science and Technology's State Response to "Rising Above the Gathering Storm" Education Task Force.

**Marlena Jones** (Panelist) is master teacher program coordinator for Math for America (MfA) DC. She helped start the Master Teacher Fellowship Program at MfA DC in 2010. Additionally, she works with the Carnegie Academy for Science Education as coordinator of programs. Jones has also taught science, technology, engineering, and mathematics and laboratory science at the college and secondary school levels.

**Jay Labov** (Moderator, Convocation PI) is senior advisor for education and communication for the National Research Council and National Academy of Sciences. He has directed or contributed to 24 National Academies reports and has served as director of committees on K-12 and undergraduate science education and the National Academies' Teacher Advisory Council.

**Susanna Loeb** (Committee Member) is the Barnett family professor of education at Stanford University, faculty director of the Center for Education Policy Analysis, and a co-director of Policy Analysis for California Education. She specializes in the economics of education and the relationship between schools and federal, state, and local policies.

**Steve Long** (Panelist) currently teaches pre-advanced placement chemistry and *ChemCom* at Rogers High School (RHS) in Rogers, Arkansas. He has served as the RHS science department chair and the Rogers Public Schools secondary science curriculum specialist. In his 39-year career, Long has taught chemistry, biology, earth science, and life science. He served as chair of the National Academy of Sciences Teacher Advisory Council.

**Margo M. Murphy** (Moderator) is a teacher at Camden Hills Regional High School, Rockport, Maine. She currently teaches freshman integrated science and junior-level botany, and she also serves as a technology integration team leader. She taught science for 22 years at Georges Valley High School where she served as department chair, K-12 science team facilitator, and in several other roles.

**Nafeesa Owens** (Panelist) serves as a program director in the Excellence Awards in Science and Engineering Program at the National Science Foundation (NSF). Prior to joining NSF, she was an associate at Quality Education for Minorities Network. She also served as director of the California State Summer School for Mathematics and Science at the University of California, Santa Cruz, where she oversaw summer programs for middle and high school students.

**Anthonette Peña** (Panelist) is the program director for the Albert Einstein Distinguished Educator Fellowship Program at the Triangle Coalition for STEM Education. Throughout her career, Peña has worked to increase the number of underrepresented groups within the science, technology, engineering, and mathematics (STEM) pipeline and to encourage students to consider STEM careers. She was an eighth-grade science teacher in DC Public Schools and an Einstein fellow.

**Steven Robinson** (Committee Member, Panelist) is in his first year at Democracy Prep Charter High School in New York City, where he teaches physics, honors physics, and senior science seminar. He has also taught in Oregon at both the high school and the college levels, and at the University of Massachusetts where he was an assistant professor of biology. He served as a special assistant, White House Domestic Policy Council.

**Heidi Schweingruber** is director of the Board on Science Education (BOSE) at the National Research Council (NRC). She oversees and helps develop many of the projects in the BOSE portfolio. Prior to joining the NRC, Schweingruber worked as a senior research associate at the Institute of Education Sciences in the U.S. Department of Education.

**Terri Taylor** (Panelist) is the assistant director for K-12 Education at the American Chemical Society (ACS). Taylor is currently working on the team responsible for launching the American Association of Chemistry Teachers in September 2014. Prior to joining ACS, she taught high school and community college chemistry in the Baltimore and Washington, DC, metropolitan areas.

**Mike Town** (Panelist, Planning Committee Chair) is a teacher at Tesla/STEM High School in Redmond, Washington. In his 29 years of teaching, he has taught algebra, astronomy, biology, chemistry, forest ecology, and horticulture. Town has served as an Einstein fellow, working as a science, technology, engineering, and mathematics (STEM) education policy analyst with the National Science Board. In addition, he co-coordinated a report from 32 Einstein fellows on how the National Science Foundation could meet STEM education recommendations.

**Suzanne M. Wilson** (Presenter) is a Neag endowed professor of teacher education, Department of Curriculum and Instruction at the University of Connecticut. Previously, she was a university distinguished professor at Michigan State University. She is co-principal investigator on Learning Science Through Inquiry with the Urban Advantage: Formal and Informal Collaborations to Increase Science Literacy and Student Learning. She also chairs the National Research Council Committee on Strengthening Science Education Through a Teacher Learning Continuum.